Knowing everything about nothing

T0296942

Knowing everything
about nothing

Specialization and change in scientific careers

JOHN ZIMAN

Department of Social and Economic Studies, Imperial College,
of Science and Technology, London

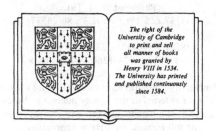

The right of the
University of Cambridge
to print and sell
all manner of books
was granted by
Henry VIII in 1534.
The University has printed
and published continuously
since 1584.

CAMBRIDGE UNIVERSITY PRESS
Cambridge
New York   New Rochelle
Melbourne   Sydney

CAMBRIDGE UNIVERSITY PRESS
Cambridge, New York, Melbourne, Madrid, Cape Town, Singapore,
São Paulo, Delhi, Dubai, Tokyo

Cambridge University Press
The Edinburgh Building, Cambridge CB2 8RU, UK

Published in the United States of America by Cambridge University Press, New York

www.cambridge.org
Information on this title: www.cambridge.org/9780521126076

© Cambridge University Press 1987

This publication is in copyright. Subject to statutory exception
and to the provisions of relevant collective licensing agreements,
no reproduction of any part may take place without the written
permission of Cambridge University Press.

First published 1987
This digitally printed version 2009

A catalogue record for this publication is available from the British Library

Library of Congress Cataloguing in Publication data
Ziman, J. M. (John M.), 1925–
Knowing everything about nothing.
Bibliography
Includes index.
1. Scientists – Vocational guidance.   2. Specialism
(Philosophy)   I. Title.
Q147.Z55 1987   502.3   86-26377

ISBN 978-0-521-32385-7 Hardback
ISBN 978-0-521-12607-6 Paperback

Cambridge University Press has no responsibility for the persistence or
accuracy of URLs for external or third-party internet websites referred to in
this publication, and does not guarantee that any content on such websites is,
or will remain, accurate or appropriate.

'*A* philosopher *is a person who knows less and less about more and more, until he knows nothing about everything.*

A scientist *is a person who knows more and more about less and less, until he knows everything about nothing.*'

Contents

Introduction

In the autumn of 1980, the late Professor Frank Bradbury, Coordinator for the Joint Committee of the Science Research Council and the Social Science Research Council, introduced me to Mr (now Professor) R.J.H.Beverton FRS. Ray Beverton had recently retired from 15 years as Secretary and Chief Executive of the Natural Environmental Research Council where he had had to grapple with the following managerial problem: For reasons of government policy, certain research establishments had had to be radically re-organized and re-oriented towards new problems. But a number of members of the permanent staff of these establishments strongly resisted any attempt to move them to different jobs within the Research Council. This was understandable where it involved moving house to a distant part of the country. But many of them were also resisting any substantial change in the subject of their research, even when this lay clearly within their capabilities.

This was puzzling. Scientists are supposed to be go-ahead people, who welcome novelty. Why should they object to doing research in a field which they had not worked in previously, but for which they seemed well qualified by their training and experience? What was it about their work, or their professional careers, or the way they were employed, that made them so rigidly specialized and resistant to change? Was there a stage in their careers when they became so unadaptable that it was scarcely worth the managerial effort to redeploy them on new problems when they were no longer needed in their previous work?

This issue intrigued me. I had always been interested in the personal aspects of the scientific life — the way in which scientists are educated, trained to do research, drawn into the scientific community, and eventually win recognition for their achievements. As far back as 1960 I had written a piece on the emergence of research as a profession (Ziman 1981), and in later years had even made two separate but abortive attempts to write a general book on science as a career. Ray was asking a practical question that touched upon this subject at every point. Other senior research managers were also very concerned about this matter, and in May 1981 (aided by Mr Harold Palmer, who had taken over as Coordinator for the Joint

Committee) we were delighted to get a one-year grant from the SRC to start work on it together.

In a sense, it was a very familiar problem − yet surprisingly unexplored. The high degree of specialization in modern science is notorious. Sociologists of science have made extensive studies of research specialties − yet they have done almost nothing on scientists as specialists. Everybody knows that scientists become very attached to their subjects, and are often very reluctant to take up problems in other fields − yet there is very little in the literature on research management on how to deal with the personal problems that arise when they get into a rut, or their work gets out of date, or their services are needed for new research programmes.

After a brief library search, we came to the conclusion that this was not one of those practical issues on which there already existed a coherent body of applicable theory. In the past five years, I have had the pleasure and profit of discussing this issue with Lotte Bailyn (MIT), Barry Barnes (Edinburgh), Stewart Blume (Amsterdam), Daryl Chubin (Georgia Tech.), Thomas Gieryn (Indiana), Gerard Lemaine (Paris), Dorothy Griffiths (London), Tom Kitwood (Bradford), Nigel Nicholson (Sheffield), Terry Shinn (Paris) and other social scientists whose published research had contributed significantly to my understanding of various aspects of the central problem, and they all confirmed that it had not previously been studied systematically.

Since social theory was not very helpful, we turned to social practice. Through the summer of 1981, we discussed the whole matter informally and confidentially with a number of heads of research councils, departmental chief scientists, directors of research laboratories, *et al*. These interviews were taped for our own convenience of retrieval, but were not transcribed, and are not quoted verbatim in this book. Since then, I have had similar talks on the same topics with very senior scientists in the United States, France and Holland. I will not list our informants by name, for they might not now hold anything like the same opinions, nor wish to be associated with my present conclusions. Nevertheless, I am extremely grateful for the time and thoughtful attention they gave us.

These discussions with 'the great and the good' certainly opened up the subject for us very effectively. From them, we could sketch out and interrelate its salient features, and determine the lines along which our further enquiries might proceed. We began to appreciate some of the issues that really needed to be clarified, such as the difference between 'versatility' and 'adaptability', the influence of promotion procedures on specialization, the role of the 'scientific generalist' and so on.

Nevertheless, for all their wisdom and experience, these eminent and influential scientists were no longer in a position to see the problems of career change from the viewpoint of a person of more modest talent or standing. In many cases, they owed their present authority precisely to having made such changes successfully several times in their own careers, and found it difficult to appreciate the unease and fear of failure of someone faced with this possibility for the first time.

To find out what ordinary working scientists thought about the whole subject, we arranged a series of small informal meetings at various research establishments. At each of these meetings, we put a tape recorder on the table, and started off a discussion ranging over all the issues that we had identified. This book is based primarily on these group discussions. Every phrase, sentence, or paragraph (whether or not 'displayed' in the text) that appears in quotation marks without specific attribution comes from this source.

The context of each discussion was very straightforward. From September 1981 to April 1982, we visited 15 research establishments, of which seven were run by research councils, five were in the public sector, two were in private industry, and one was part of an academic institution. Each visit was arranged for us by the director of the establishment, who was carefully advised of the purpose of the meeting. A group (in three cases, two separate groups) of about eight members of the permanent research staff of the establishment would be introduced to us, with the clear understanding that this was not a managerial investigation, and that nothing they said would be reported back or could ever be attributed to them in the final report of the research. Almost all the discussions were, in fact, very free and easy, and apparently uninhibited, with frequent references to the failures and follies of their superiors. This was corroborated by the noticeable change in tone on the only occasion when a senior manager of the establishment took part in the discussions — and then withdrew.

We had no detailed control of the choice of participants. Sometimes they were all members of a single research group, and knew each other well; sometimes they came from different research groups within a large establishment and only knew each other by sight. Although we felt that they were reasonably typical of the communities from which they were drawn, they could not be supposed to be statistically representative of that population. Since participation was voluntary, it is possible that these were amongst the more successful, more articulate and less narrowly specialized members of the staff of the establishment. They ranged in age from the late 20s to the middle 50s, but the median age was 39, and more than one third of them were between 36 and 40. In the terminology of the scientific civil service, they varied in rank from HSO to SPSO (see §3.3), but about half of them were in the 'career grade' of PSO. Most of the participants could thus be described quite fairly as being 'in mid-career': they were already solidly established professionally, but could still see themselves as working hard at their jobs for many more years.

The discussions were as open-ended and unforced as we could make them. Each participant had been given a two-page account of the purpose of the project, which was explained again, informally, at the beginning of the session. Then, to start things off, we would ask one of the participants to tell us about his or her past career — what courses they did at university, what jobs they had then taken, how they had come into the establishment, what projects they had worked on, and so on. This would soon bring up one of the topics in which we were interested, which could then be turned over to the group as a whole for general discussion. As a matter of courtesy, we would ensure that each member of the group had had an opportunity to say something about their own career: in fact, since they often did not know each other well, this was of interest to everybody else!

By the time we had gone round the whole group in this way, we would have been offered opinions on most of the topics on our private check list without having had to pose them as explicit questions. Sometimes, of course, the conversation would flag, and we would have to prime it with an outright question, or probe for a positive opinion. Generally speaking, however, they did not need much prompting, and a couple of hours would pass in quite lively discussion, approximating in tone to what such a group might say amongst themselves over coffee or lunch. Because both Ray and I were known to be professional scientists — albeit older and of higher status — we could join in this conversation in a natural way. At the end, when we thanked them for their co-operation, they would often say how interesting it had been to talk about these matters, and how much they had enjoyed it.

The record of these discussions occupies some 50 hours of tape. It constitutes a raw sample of characteristic public discourse on a matter of personal concern to each of the participants. One may assume that it contains references to most of the considerations that members of the group deem relevant, conceptualized and formulated in the manner that they deem intelligible and acceptable to other members. In other words, it samples the common experiences and commonly held notions of people of that kind, in so far as they are willing to reveal them to their acquaintances.

These tapes have obvious limitations as a basis for a systematic analysis of the problem. Because the participants in the discussions were not a statistically representative sample, no quantitative indicators can be derived from them. Because the questions that we had in mind were not directly posed as such to the participants, the answers that we got are often contradictory, or inconclusive, and cannot be structured into logically coherent sets. And because each discussant spoke in the presence of colleagues or acquaintances, he or she must have concealed many private hopes or fears which might have been confessed in confidence to a sympathetic interviewer.

It is doubtful, however, whether a different research methodology would have produced more useful data at this stage of the investigation. It would not have been possible to tap a statistically representative sample of the relevant population except through some impersonal written communication which could be dealt with quickly by relatively uninterested informants. A thorough survey by questionnaire would have given quantitative answers to specific factual questions, such as the prevalence of career changes of various types, but only at the expense of imposing a preconceived categorical framework on more subtle aspects of the matter. Individual interviews with selected informants would have probed deeper, but would have sampled much less widely in the time available, and we doubted whether we had the professional skill to carry out such interviews without projecting our own viewpoint on our informants. These and many other considerations are, of course, familiar methodological issues in the social sciences.

Our original plan, in fact, was to use the group discussions as a means of exploring in detail a very complex landscape which had never been mapped from this point of view. We hoped at first to compare the attitudes and experiences of British scientists with their contemporaries in other countries, and I did have the opportunity, whilst visiting the United States on other business, to record similar group discussions in two major industrial research laboratories. We then intended to use other methods, such as questionnaires and interviews, to study further any critical points that might emerge. It soon became obvious, for example, that the procedures for Individual Merit Promotion (§7.6) play a very important part in career patterns in the Scientific Civil Service and the research councils, and ought to be looked at much more closely. We even promised some of our discussion groups that we would come back to them and try out our preliminary conclusions on them, to correct our impressions and obtain further insights.

This promise could not be kept. Perhaps our plan was too ambitious. In any case, circumstances prevented us from carrying it out. At the end of April 1982, Ray Beverton took up an important international assignment which prevented his further regular participation in the project, and I decided not to apply for a continuation of the SRC grant, at least until I had made a systematic analysis of the existing raw data. As the date of this publication indicates, this took much longer than I expected. Perhaps a year of this delay was due to the unanticipated demands of other work, but the task I had set myself was surprisingly laborious and the present text is substantially longer than I originally thought likely.

The first step was to have the tapes transcribed for detailed scrutiny. For this I am extremely grateful to Lilian Murphy and Felicity Hanley, of the H. H. Wills Physics Laboratory of the University of Bristol, where I then held an appointment. It was a lengthy job, which had to be fitted in amongst their other duties, and they often had to cope with extraneous noises, poor diction, and people talking simultaneously. Nevertheless, within six months they had produced a complete text of all the discussions, amounting to the equivalent of about 750 pages of single-spaced typescript.

The next step was to go through the transcripts, marking relevant passages and indexing them according to a preliminary scheme based upon the interim report that we had already prepared for the Science Research Council. This index was then analysed and restructured until it began to look like a tentative synopsis for the present book. Finally, as I came to each section, I retrieved the relevant passages from the file of transcripts, selected parts of them for quotation, and linked them together into a coherent text.

This book consists in large part, therefore, of the actual words of working scientists about their careers. But I have not followed current sociological fashion by treating the verbatim text as if it were sacrosanct. For brevity, the quotations have been pruned of minor irrelevancies, repetitions, and vacuous phrases such as 'I think', 'I mean', etc. To preserve the anonymity of our informants, all personal names have been excised, and words that might identify a particular establishment or scientific specialty have been generalized. For clarity, the grammatical hiccups of impromptu speech have been tidied up in accordance with the obvious intentions of the speaker, or gaps filled with appropriate link words. All such changes from the transcribed text are indicated by ellipses . . . or by brackets []. In my opinion, the information thus lost is insignificant by comparison with the unavoidable effects of background noise and ambiguities of transcription, let alone the elimination of information about the speaker and the context of the conversation from which the particular quotation had been selected. But when in doubt whether or not to quote a particular passage, I would usually decide to include it, if only to indicate how diverse and contradictory people's attitudes can be.

Perhaps the most serious defect of this material is that it is now five years out of date. The project was triggered off by the general feeling that British science was going through a grave crisis of confidence. As it turns out, what seemed like a sorry state then has got worse. Many threatened cuts and closures have proved even more severe than was then feared, some establishments have been completely reorganized, and most research programmes have been radically reorientated. As a consequence, the public morale of the British scientific community is much lower now than it was even five years ago. But that does not mean that scientists as individuals have not been able to cope with these rapid changes in their situation. As this study shows, they are usually more versatile than they tend to believe, and adversity may even have strengthened them professionally by forcing them to become more flexible and adaptable. Nevertheless, it would be immensely instructive to go back again to these establishments, as we half promised, and look for changes in attitudes towards change itself.

The overall structure of the book is as follows: Chapter one is about scientists as subject specialists. Chapter two is a schematic account of the organizations that employ scientists to do research. Chapter three deals broadly with the careers of scientists in organizations that are now undergoing rapid change. In Chapters four and five I discuss the practicalities of a personal change of research specialty and the motives that a scientist might have in resisting or welcoming such a change. Chapter six is concerned with career changes *out* of research, into management, administration, etc. Chapter seven then goes through various organizational policies and practices which seem to have some effect on the versatility and adaptability of scientists, and Chapter eight suggests a number of practical steps that can be taken to help each individual through periods of career change. In the final chapter I have tried to set the earlier conclusions into a wider setting of national science policy, of international comparisons, and of the changing role of the 'scientist' in modern society. I suppose it all adds up, in the end, to asserting that scientists are really much more versatile and adaptable than they or other people tend to think, and it is to everybody's advantage to give them the time, the opportunity, and the sympathetic leadership to face the challenges of radical change. But there can be no better way of convincing you of this than by inviting you to read at length the evidence presented on the following pages.

In thus trying to account for the genesis and structure of this book, I have already mentioned by name a number of people to whom I am particularly indebted for advice and assistance. To these should be added the 100 or so working scientists who took part so enthusiastically in our discussions, and the directors and administrative staffs of some 15 research establishments who arranged these meetings for us. I am grateful to Nicola Kingsley for the preparation of this index. And one of the real benefits of the whole project was spending many days in the company of Ray Beverton, driving up and down the country, meeting people together, and discussing every aspect of the scientific life to which both of us had been so long committed.

Imperial College
London
April 1986

1
Research Trails

1.1 Maps of Knowledge

By its very nature, scientific work is minutely specialized. Scientists often compare themselves to masons, each adding a few tiny bricks to the vast edifice of human knowledge. Science grows by a systematic division of labour, where the domain of action of each worker is narrowly limited and the work to be done within each domain is highly skilled. The practical expertise and understanding needed to undertake a serious scientific investigation usually takes many years to acquire. However much we may deplore it, almost all research nowadays has to be done by specialists.

All professions are to some degree specialized. A lawyer may tend to specialize in cases of a particular kind, and become known as an expert on, say, patents, or marine insurance. An architect may gain a reputation for designing schools, or churches, or museums. In any substantial enterprise, such as a coal mine, or a steel works, there will be people known to have very specific skills which are essential to the enterprise and which may have taken half a lifetime to learn. But no other occupation is so finely and distinctly subdivided into 'specialties' as science.

What do we mean by a scientific specialty? As in other occupations, individual scientists bring to their work a wide range of skills acquired by personal experience. Most of these skills are *tacit* (Polanyi 1958); they cannot be defined precisely, or catalogued systematically. But science differs from other highly technical activities in that its goal is the production of organized knowledge (Ziman 1984). As every scientist knows, a research report is unacceptable unless it cites previous work on the subject. This is not just a convention. Philosophers are now beginning to realize that a genuine scientific question cannot even be formulated without reference to a background of existing knowledge (de Mey 1982). Whatever other skills may be needed, a researcher must be more or less familiar with what is already known scientifically about the problem to be tackled. When someone is needed to undertake research on a particular subject, the natural choice is a scientist who is already well informed on that subject, whether by formal education or by previous research on closely similar problems. A high degree of

2

Knowing everything about nothing

subject specialization is thus an indispensable condition for progress in every field of science.

From this point of view, any distinct subdivision of scientific knowledge may be considered a subject specialty. It is the extreme narrowness and specificity with which such knowledge can be so divided that differentiates scientists so minutely from one another. The very fact that science is an organized body of knowledge implies that every research project or discovery has its recognized place in a larger scheme. But there is no way of setting up such a scheme without defining distinct categories of knowledge according to some rational criteria. The information produced by research in a mature scientific discipline, such as chemistry or neurophysiology, may be classified according to a number of well-defined and precise ordering principles – general theories, species of organism, classes of compound, techniques of investigation, and so on – and can thus be subdivided into innumerable 'subjects' for specialized research.

To appreciate the depth to which scientific knowledge can be divided into distinct categories, it is only necessary to look at one of the standard classification schemes used by librarians and editors to arrange and index journal articles, research reports, books, and other scientific documents. The archetype of most such schemes is the Universal Document Classification system, where the various categories and subcategories are arranged in a hierarchy, according to the successive digits of a decimal number. The internationally approved scheme for physics, for example (ICSU 1975), assigns to each subject an alphanumeric address tag of five characters.Thus, a subject on which I used at one time to specialize is clearly defined and tagged:

72.15.C ELECTRICAL AND THERMAL CONDUCTION IN AMORPHOUS AND LIQUID METALS AND ALLOYS

But this alphanumeric tag is not arbitrary. It shows that this subject is part of a larger category:

72.15 Electronic conduction in metals and alloys

This, in turn, is a subdivision of:

72. Electronic transport in condensed matter,

which is just one of the ten subsections of:

70. CONDENSED MATTER: ELECTRONIC STRUCTURE, ELEC-TRICAL, MAGNETIC, AND OPTICAL PROPERTIES

– one of the ten 'subdisciplines' of physics as a whole (Fig. 1). In other words, physics can be broken down into several thousand 'subjects', each classified hierarchically to the fourth decimal digit.

It might be objected that physics is a peculiarly well-ordered science, because of its analytical approach to nature. But even such an empirical, practical discipline as forestry can be classified into a quasi-decimal hierarchy of 'Divisions',

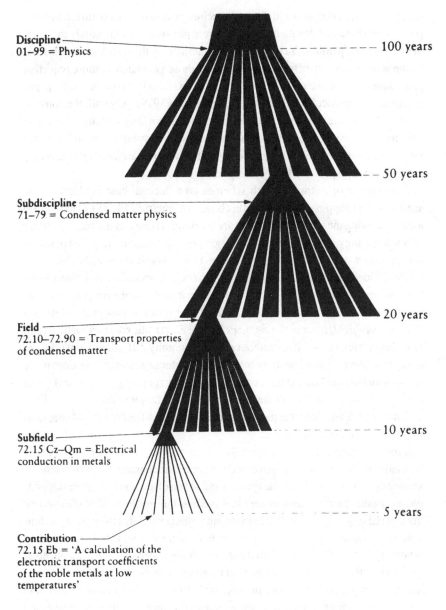

Discipline
01–99 = Physics

100 years

50 years

Subdiscipline
71–79 = Condensed matter physics

20 years

Field
72.10–72.90 = Transport properties
of condensed matter

Subfield
72.15 Cz–Qm = Electrical
conduction in metals

10 years

5 years

Contribution
72.15 Eb = 'A calculation of the
electronic transport coefficients
of the noble metals at low
temperatures'

Fig. 1

'sectorial groups', and 'working groups' on the same lines (IUFRO 1982). All such classification schemes are, of course, essentially artificial, and do not pretend to represent the actual structure of knowledge in detail. A hierarchical scheme cannot, for example, represent the numerous cross-connections corresponding,

say, to the results obtained with the same technique on a number of different com-
pounds, or to the fact that many quite different physical systems satisfy the same
mathematical equations. Yet the specialties defined by the conventional classification schemes do not differ profoundly from those generated by more objective
procedures, such as the detailed analysis of the citations from one scientific paper
to another in a particular field of research (Garfield 1979). After all, the conventional schemes are designed in consultation with working scientists, and thus
conform to the conceptions that scientists themselves have of the boundaries and
interrelations of the various subjects on which they do research and on which they
are counted as experts by their colleagues.

The practice of setting up such schemes on a decimal base is clearly quite
arbitrary. But suppose we think of this base as a 'metric', like the grid lines on
a map, showing the scale on which we are working. Using this metric, we can set
up a terminology indicating the actual degree of specialization required of scientists in their work. Thus, we may say (Fig. 1) that a typical scientific discipline
can be divided into about ten *sub-disciplines*, each of which is divided into about
ten *fields*, and so on. Indeed, as we have seen in the case of physics, the information to be classified may be so well defined that it can be assigned to one of a
thousand *sub-fields* (such as the category **72.15 : Electronic conduction in metals
and alloys** referred to above) without risk of ambiguity.* If, for example, I were
to say that 'the research specialty of a typical academic scientist is seldom more
than one or two fields in extent' I would mean (very roughly) that it would cover
only one or two per cent of all the subject matter of a major academic discipline,
and thus, by implication, that there might be 50 or 100 different sorts of specialist
in that domain of science alone.

It must be emphasized, however, that the extent to which scientists actually
specialize in their work is not governed simply by the characteristics of scientific
knowledge. As we shall see, the breadth and shape of a scientific career depends
on many other factors, such as personal inclination, institutional traditions and
managerial imperatives. All that I am saying at this point is that the primary definition of a scientific specialist is by his or her 'subject', and that this refers to a
relatively well-defined and limited segment of the enormous body of concepts,
data and methods known to science. In more metaphorical language, scientists
specialize in particular 'areas', or 'regions' of the 'map' of knowledge. These
specialty regions are not clearly delineated on this map, but they naturally tend
to conform to typical features of the scientific 'landscape' — a powerful theoretical
paradigm, say, or a particularly challenging problem, or the discoveries made

---

* This terminology follows neither Chubin (1976) nor Gieryn (1978), who use the theoretically loaded
  words 'specialty', and 'problem area' where a purely conventional term such as 'field' begs less
  of a question.

possible by a new technique. That is why it is essential to understand some of the characteristics of such landscapes, before following the trails of those who set out to explore them.

1.2 Changing scientific landscapes

Science is essentially a *progressive* enterprise: it is dedicated to *change*. This is not to say that the accumulation of scientific information and understanding is necessarily for the better: it is simply to observe that every 'map' of scientific knowledge is in constant flux. The whole notion of subject specialization must take account of this process of *cognitive change*, which is thus one of the most important factors in the life of every scientist, and a major aspect of the present study.

Every scientific specialty is being transformed continually by internal discoveries or external influences. Sometimes this transformation is slow and uninteresting, as in the accumulation of data by routine methods: at other times, an exciting new discovery opens up many old problems for solution, and suggests many new ones. In the latter case, the subject may grow rapidly, both in the amount of research being done and in its intellectual scope.

But a scientific specialty cannot expand by a large factor without rupturing the classification scheme by which it was defined. A scientific 'breakthrough' is not simply the discovery, or the opening up, of a region of the scientific landscape which has not previously been explored: it usually implies a significant change in all existing conceptions of the territory around it, and thus amounts, in effect, to a change of the 'landscape' itself. It is not enough to obtain a whole lot of new answers to old questions. If there is to be genuine progress, this information must be brought under intellectual control, and made the basis for new theories, new concepts, new techniques − and new scientific questions. Thus, for example, once it became clear that 'continental drift' was a real process, almost every fact and theory of geology had to be re-assessed and reinterpreted, and the new subject of 'plate tectonics' came into being.

The history of every science, the biography of every scientist, bears witness to the universality of radical cognitive change. We need not enter here into the scholarly debates about the way in which such 'scientific revolutions' actually occur, nor whether they represent a change to a state of knowledge that is 'incommensurable' with what it was before (see, e.g. Ziman 1984). What we can say is that, after such a revolution of thought, the way in which scientists had previously classified the objectives and results of research no longer holds good. The time has come to draw new 'maps' of the subject, and to delineate on them new specialty boundaries.

Everybody has heard of the great scientific breakthroughs that revolutionize whole disciplines. What is not always appreciated is that the same process of cognitive change occurs at every level of specialization, down to the narrowest 'subfield' of research. In my own former subject of theoretical physics, for example, research on the highly specialized topic of the electrical conductivity of liquid metals was revolutionized some 20 years ago by the discovery of a simple formula that at last made sense of most of the existing experimental data. On a much broader scale, 'the physics of condensed matter' began to be recognized as a distinct subdiscipline about 50 years ago, when it became clear that quantum theory could be applied to the behaviour of electrons in metals. But this was only one episode in the vast transformation of physics as a whole, following on the discovery of nuclear and atomic phenomena at the beginning of the century.

The same sort of process can be observed in every active domain of research. Generally speaking, the narrower the scope of a specialty, the shorter the time scale on which it will be seen to change. Citation analysis shows, for example, that the effective lifetime of a primary paper in physics is only about five years (see, e.g. Ziman 1985): if it has not made its mark by then, and been incorporated into a larger movement of change, then the chances are that it will fall into oblivion. My guess is that the lifetime of an active 'subfield' of physics is about ten years, and that 'fields' of research grow and decline in periods of the order of 20 years. A typical 'subdiscipline', such as nuclear physics, probably retains its identity for 40 or 50 years, whilst the transformation period of physics itself, considered as a major 'discipline' within the natural sciences, might be something like a century.

These estimates of the time scale of cognitive change in science are, of course, purely notional. I have chosen physics as my example because it is the only discipline that I know well enough to make such guesses. The general scale of change may be somewhat different in other sciences, and certainly varies greatly from specialty to specialty. Some major disciplines, such as cell biology, have evidently changed out of all recognition in a few decades, whilst there are numerous minor fields which have remained essentially unchanged for long periods. The main point is that the landscape of every science — and of the sciences as a whole — is being continually reconceived and remapped. The occasional grand revolutions that attract the attention of historians and philosophers do not account for all the radical change that is actually occurring in science. This change is going on all the time, at different rates on different scales.

The most obvious type of change in the landscape of science is the appearance of a new subject combining information, concepts or techniques from previously distinct domains of research. Thus, the modern discipline of molecular biology is not simply a revolutionized form of, say, the established discipline of biochemistry, but combines elements from genetics, microbiology, chemistry and physics. The development of entirely new *interdisciplinary* subjects is characteristic of the history of science, and again takes place on every scale. In fact, what might seem to outsiders as the 'internal' transformation of a discipline or subdiscipline would usually be described be described by experts as the appearance and growth of a new 'field', combining elements from previously unrelated areas of research in order to tackle new problems. This often occurs when research is deliberately focussed on some practical problem of political or economic significance. In agricultural research, for example, a new 'problem area' − that is, a potential new specialty − is created whenever the intellectual resources of several different biological disciplines are brought to bear on an urgent problem such as the prevention of a particulary serious plant disease or the improvement of the yield of a major crop.

It is important to note, however, that interdisciplinary subjects do not remain unspecialized for long. As each new region of the scientific landscape is explored and mapped, it is subdivided into 'fields' and 'subfields' just like any traditional scientific discipline. Where the subject matter is diverse, and many different approaches are possible, it may not be easy to place these new specialties on a simple hierarchical classification, or even to represent their relationships by adjacency on a two-dimensional diagram. Nevertheless, as we saw in the case of forestry (§1.1), experts recognize the existence of such categories, and are accustomed to defining themselves and each other in relation to them.

1.3 The scope of a personal specialty

Research scientists are individually credited with special expertise over particular domains of scientific knowledge. These *personal* specialties are defined by reference to current 'maps' of knowledge, but they are not actually specified on such maps. Inspection of the subject classification scheme of a discipline, for example, will give an idea of the problems, concepts, techniques, etc. on which such personal specialties might be centred, but it will not indicate their position or scope. In fact, as we shall soon see, the latter is itself a highly individual trait: scientists not only locate themselves very unsystematically over the scientific 'landscape'; they also vary greatly in the area of science over which they are accorded or claim special expertise.

This individual diversity is vital in all that follows. The present investigation would be pointless if the division of labour in science were strictly formalized and controlled. Research work is not divided like clinical medicine into *professional* specialties, where practitioners must take examinations to qualify as 'specialists' in particular branches of the subject (§3.1). Many stiff examination hurdles may bar the way into a research career, but once these are cleared there are no official limitations on the subjects on which a scientist with a degree in a particular discipline might in principle carry out research (§4.2). It might be foolish for a PhD in botany to try to make observations in astrophysics, but it would not be illegal to do so. For this reason, there is little to be learnt from data on occupational specialization (e.g. Fiorito 1981) based solely on academic degrees and other official professional qualifications.

A much finer-grained classification is needed to demonstrate the actual diversity and specificity of work in research. For example, in the manpower surveys carried out by the American Institute of Physics (Porter 1975, 1976), more than ten 'employment specialties' were distinguished within the single subdiscipline of *nuclear physics*, and more than 20 in the subdiscipline of *optics*. In other words, these surveys suggest that a working scientist usually has no difficulty in defining an area of personal expertise on the scale of a research 'field', such as *nuclear spectroscopy* or *atmospheric and space optics*, whose scope might be less than one per cent of the extent of a conventional 'discipline' such as *physics* or *electronic engineering*.

This estimate of the actual extent of specialization in scientific work is confirmed by direct study of the range of subjects covered in the output of individual research scientists. Taking a standard subject classification scheme as his starting point, Gieryn (1979) found that research in astronomy could be classified into about a hundred 'problem areas'. Astronomy may be considered an academic discipline in its own right, but it is not so large as physics or chemistry, so that each of these is probably somewhat smaller than what I would call a typical research 'field'. Gieryn then looked at the papers published by more than 2000 American astronomers over the period 1973–5, and found, in the majority of cases, that the work of each astronomer was concentrated into just two or three such areas. In other words, judging from his or her published work, the personal specialty of a typical astronomer would appear to be about one 'field' in extent at any one time.

The terms that scientists themselves use to describe their personal research specialties are consistent with this estimate of their size. A hydrologist remarks that at one time he was 'doing chemistry of sea water, chemical oceanography', and that this was quite different from the chemistry of river water. Research on the endocrine function in fish is clearly differentiated from human endocrinology, even though the underlying physiological mechanisms are thought to be closely related. An experimental physicist reports that all of his research up to a certain date had been at very low temperatures:

 '. . . most of my life it had been below 4° and the last five years below ten millidegrees . . . '.

Table 1.1

Number of problem areas	Cumulative percentage
1	20
2	40
3	60
4	70
5	80
6-7	90
8+	100

Most working scientists evidently see themselves as occupying only one or two of the thousands of 'fields' into which the labour of the scientific enterprise is divided.

Any quantitative estimate of the size of the typical unit of subject specialization in science is obviously highly suspect, and should not be accorded the intellectual dignity of arithmetical manipulation. But such a unit could not be very much larger than a 'field' simply because of the enormous rate at which scientific knowledge accumulates. Remember that something like 100,000 scientific papers are published in a major scientific discipline each year. To keep up with the literature in a single research 'field' one would need to be aware of the contents of one per cent of these — that is, of a thousand original papers. This task is not quite as heavy as it seems, since the quality of the scientific literature is so uneven that one can safely ignore at least 90 per cent of it. Nevertheless, to remain perfectly informed over the whole of such a 'field' one would have to become acquainted with the actual research results of something like a hundred papers each year, and arrive at some personal opinion as to their validity. A single 'field' thus provides plenty of material on which to become a 'subject specialist', without looking further afield.

This is not to suggest that every scientist has to be a world authority on his or her subject in order to do good research. In practice, some scientists do try to follow this counsel of perfection, and work only in very narrow areas, whilst others have wider interests. This comes out clearly in Gieryn's study of the publication records of astronomers. For example, the cumulative percentages of the sample who had published at least one paper in each of several distinct 'problem areas' are shown in Table 1.1. Thus, although the majority of the scientists in this group seemed to be working in what would be, by our present reckoning, just one or two 'fields', there were a significant proportion who did not appear to be nearly so specialized in their research. It is interesting to note, moreover, that the publications of each of these more 'diversified' scientists were not always concentrated into a single broader category of the subject classification scheme (e.g. into what we would call a single 'subdiscipline'), but were sometimes distributed quite widely over astronomy as a whole. In the absence of further detailed evidence on this point, we should not give too much weight to the data of Table 1.1, but they do give a rough indication of the variability, from individual to individual, of the extent of subject specialization in an active domain of scientific research.

1.4 Persistence

At any given moment, a scientist may say that he or she is a specialist in a particular 'field' − *the behaviour of aphids on cereal crops*, for example, or *computer modelling of airflow at high Mach numbers*. But a particular specialty is not necessarily a permanent personal attribute. A few years later, the same scientists might report that they were now interested mainly in, say, *the application of pesticides to fruit trees*, or *the formation of droplets in fluid jets*. Quite apart from the changes that naturally occur in the definition and classification of scientific 'subjects' (§1.2) an individual scientist's personal specialty may change radically in the course of his or her career. In fact, the circumstances and consequences of this familiar feature of scientific life are the main theme of this book.

But before we enter into a detailed discussion of these circumstances and consequences, we need to have some quantitative notion of the scale of this phenomenon? To what extent, and in what manner, do researchers actually change their 'subjects' during their scientific careers? These are straightforward questions, which could be answered by direct analysis of the published work of a number of mature scientists in a particular area of research. Such a collection of personal *curricula vitae* would not tell the whole story of specialization and change in contemporary scientific careers, but it would indicate the characteristic patterns of such change. The *research trails* that individual scientists actually follow across the cognitive landscape (Chubin & Connolly 1982) must be considered the primary evidence in any study of the way in which they respond to the peculiar demands of their vocation.

Unfortunately, there is little systematic information on this subject. Excellent data for such an investigation are undoubtedly available in the personnel files of major scientific institutions, but they have not been studied from this point of view. Once more, all that we have to go on is Gieryn's analysis of the published work of American astronomers over the period 1950−75 (Gieryn 1978, 1979). Since this study covered only one relatively narrow discipline, it cannot be relied on as a quantitative account of career patterns in all scientific work, but it does give a good idea of the prevalence of various types of research trail in this particular, rather 'academic' branch of science.

Table 1.2

| Period | Percentage continuing in same | |
Years	'problem area'	'speciality' (? 'field' ?)
5	50	65
10	25	45
20	15	30

What does come out clearly is that the rate of change of personal specialties is very slow. A scientist who has once published a paper on a particular specialized subject is very likely to go on publishing on the same subject for many years. Indeed, as Gieryn points out, the longer a subject has figured in an astronomer's 'problem set', the higher the probability that he or she will continue research on that subject for a further period. From his estimates of the prevalence of this effect for a few selected 'problem areas' and 'specialties' we can deduce the actual length of time for which a scientist would normally continue to be a specialist on a particular subject. The figures given in Table 1.2 are obviously very approximate, but they clearly indicate that a significant proportion of scientists – something like 25per cent – continue to work actively in the same 'field' of research for periods of one or two decades – that is, for the major part of their professional careers.

These figures may, in fact, underestimate the degree to which, once they are well under way, scientists contine in their research specialties. According to Mullins (1972), the mean length of membership of the *phage group* was only about three years, but doubles to six years if those who published only one paper on the subject are excluded. This is consistent with what we found in our discussions. A scientist in his early 40s reported having spent ten years on essentially the same subject, and expected to continue working 'in much the same area' for a further ten years. An older scientist described the study of the mode of biological action of a particular compound as his 'life project'. Another described the problem that he had studied for his PhD, some 20 years before, and said, typically:

'Anyway, I've more or less stayed in that field ever since. I suppose it's interesting as a field – to you, that is – because it's very restricted; there must be only half a dozen people in the country who are just in that field.'

This *persistence* in their specialties is particularly noticeable in the careers of very successful researchers, especially in academia. About 50per cent of the publication records of a hundred recently deceased Fellows of the Royal Society (Ziman 1981) followed roughly the same pattern: they stayed in the same narrow specialty throughout their lives. As the biographical memoir might report (in a fictitious case), 'Bloggs did his PhD under the late Professor Coggs, who put him on to the subject of the classification of marine worms . . . . Now, with the death of Bloggs, the world has lost its greatest expert on − marine worms!'. As we shall see, lifelong specialization in a single 'field' is not the only research trail leading to success in science, but it is a career pattern that is widely prevalent throughout the scientific world.

1.5   Diversification

The stereotype of the scientist as a lifelong specialist on a very narrow subject is thus not unjustified. Nevertheless a substantial proportion of all scientific careers follow a rather different pattern. As we have seen (§1.3), about 20per cent of the astronomers in Gieryn's study seemed to be working simultaneously in five or more 'problem areas', often scattered, apparently at random, over astronomy as a whole. Something like 25per cent of recently deceased Fellows of the Royal Society had published papers on a wide variety of subjects, distributed over one or more subdisciplines of their major discipline. Our imaginary colleague, Professor Bloggs, for example, might have gone on from his early studies of marine worms to make contributions to many different aspects of marine biology, and died renowned as the world authority on the oceanic fauna.

   This tendency towards *diversification* is not confined to scientists of outstanding ability. As one of our interviewees said of himself:

   'I am, I think, a natural broadener. I heard someone say I was a butterfly, in a sense that I very early acquired interests over a very broad field, and that is the way I keep a lot of balls in the air. So I think I am a natural for being broad; not having too much penetration.'
Another explained that he had not got a broad research philosophy:

   'I don't see a single theme linking all my work together. It's just a series of interests, really. I've just said, 'Right now I would like to know how that works, or how these organisms react, or the effect of this on that sort of particular animal', and so on. And, you know, the funds have been there; something − the topic − has cropped up which has been within my general field of interest: ah, it's a good topic, that I am interested in, and I get involved in it.'
The practical rationale of diversification comes out clearly in another discussion:

   'And the, I suppose, switches in direction have been largely brought about by perhaps more interesting side lines, so I have dropped the main line of work and branched off. And also, when I have hit blocks because of lack of equipment and it's not possible to pursue the line of work I was doing, and because the equipment was not available here, so I stopped one line and started another . . . .'

'[I had] three or four distinct lines of work going on at the same time, and I would spend a block on one subject and a block on another, and perhaps on another, and then go back to A again. But in some cases I sort of terminated a branch'. It's essential, I think, to have a number of different things to do . . . It's like a system is more stable when it's more diverse. If something happens to one line of research, like somebody gazumps you, you can always − you always have a second string.'₁

And ideas trigger each other up as well. Techniques you develop in one line of research, suddenly you realize that they are applicable to something completely different, and you jump across and follow that up.'
This characteristic opportunism of much scientific work (Knorr−Cetina 1981) is to be contrasted, however, with the personal research policies of other scientists in the same group:

'My own experience is that I had to keep roughly to the same line and resist going off outside. That is, otherwise the long-term strategy gets lost.'
. . . . I can only do one thing at a time . . . . And in fact that is really true . . . . It's been my reaction in every situation that part of this problem is really understanding it well, and part of understanding it really well is going back, right back to square one . . . I find that is the way I work, and then if I feel it's time to stop . . . . I am stuck . . . I forget about it.'
In dealing with specialization in science, it is clearly essential to allow for the wide range of personal research policies amongst scientists of all levels of ability, from the narrowly convergent specialist to the highly divergent 'diversifier'.

1.6 Migration

What happens when a scientist decides, or is obliged, to stop research on a particular topic and begins research on a rather different one thus depends very much on the general pattern of his or her career. Scientists who are not highly specialized in their interests can obviously make changes in the subjects on which they do research without sharp discontinuities in their research trails. When they take up a new topic, they do not have to drop all their old research problems, but can allow quite diverse investigations to run at the same time, sometimes for years on end. Instead of looking like a narrow path with occasional sharp discontinuities, the research trail broadens into a number of tracks: forking, diverging, sometimes getting nowhere, sometimes even converging and recombining. In such a career, the first steps in a new specialty, or the last contributions to an old one, need not stand out as dramatic events.

The great majority of scientists — perhaps 75per cent — would not seem sufficiently diversified in their interests to make such changes quite without incident. Yet most scientists do in fact change their research specialties substantially in the course of their lives. Even a scientist who has persisted for a long time in a particular 'field' may eventually give it up and undertake research on a rather different subject. The important question is: do such changes occur abruptly, or are they usually spread over a number of years?

Interspecialty *migration* is often considered to play a major role in the growth of new areas of science (Chubin 1976), and reference is often made to 'a few exceptional scientists' who 'skim the milk' by moving quickly into 'new and promising areas, making several important contributions, and then moving elsewhere as competition increases and the level of significance diminishes' (Mulkay 1974). But the anecdotal literature probably exaggerates the influence of this small, atypical group (Hufbauer 1978), whose careers are envied rather than emulated by other scientists. The 'diversified' scientist, with many fingers in many pies at the same time, is much more typical.

A somewhat different pattern of 'migration' shows up in the obituaries of Fellows of the Royal Society (Ziman 1981). In the course of their lives, about 25per cent of them moved from one quite narrow specialty to another equally narrow one that might be quite a distance away on the cognitive map. Thus, for example, our fictitious Dr Bloggs might, at the age of 40, notice some peculiar enzyme in the blood of one of his precious worms, and finish up as the world expert on some recondite topic in enzymology. But evidence from this source must be treated with caution, since these were all, by definition, very succesful scientists, with very long and active careers in research. In any case, the apparent abruptness of many of these moves may be largely due to the fact that this was a generation of scientists whose research trails were severely dislocated, in mid-career, by the Second World War.

The careers of the astronomers studied by Gieryn (1978) would seem to be more typical of scientific careers in general. On a strict interpretation of the metaphor, the number of these who 'migrated' was relatively small. Thus, from 1963−5 to 1973−5, only about 10per cent of Gieryn's sample completely changed their 'problem set', and were doing research in entirely different 'problem areas' from those in which they had been working ten years before. This figure is not, of course, inconsistent with the evidence that as many as 25per cent of successful scientists completely change their specialties in the course of their whole careers; but when such a change is spread over ten years it can scarcely be regarded as abrupt. Astronomers whose 'problem set' was very narrow did tend to move to new specialties quite frequently — around 20per cent of the sample in as short a

Table 1.3

Type of change	Problem set		Percent
	1963 – 5	1973 – 5	
Duplication	a b	a b	10
Accretion	a b	a b c	35
Subsitution	a b	a c	20
Migration	a b	c d	10
Disengagement	a b	a	15
Witdrawal	a b	–	10

period as two years – but most of these seem to have been 'younger scientists whose published research is neither plentiful nor of high quality'.

All in all, the available evidence does not seriously contradict Gieryn's conclusion that out and out 'migration' is 'relatively rare' in the careers of mature scientists. Under normal conditions, they seldom move spontaneously to radically different areas of research, leaving all their previous research interests behind them. This is borne out by the interview material: in almost every case, scientists attributed any abrupt discontinuities in their research trails to the pressure of external circumstances (e.g. having to get a new job) rather than to their own personal decision to make such a change.

1.7 Drift

The fact that most scientists actually change their research subjects quite gradually can be inferred from Table 1.3 (Gieryn 1978). These data refer to more than a hundred astronomers who were each working in *two* distinct 'problem areas' ('a' and 'b', say) in 1963 – 5. Ten years later, in the period 1973 – 5, about 80 per cent were still publishing papers in at least one of these areas, but their 'problem sets' had changed in various ways. In particular, more than half of them had taken up a new specialty, either in place of a previous interest ('substitution') or as an addition to their existing set ('accretion').

These figures cover such a diversity of people and circumstances that they cannot be interpreted in detail. But the interview material shows that even those scientists who might be regarded as highly specialized are able to follow the strategy of the 'diversifiers' and weave a succession of long but finite strands of research into a continuous thread:

'In my case it just tended to evolve steadily but slowly over a period of time . . . . I haven't suddenly stopped one job . . . to start another one. I've

been in the desert for a while, but it's rather a question of running down one
area and starting another one, so that you can overlap the period of working
into one subject while finishing off results in [the previous one].'
Sometimes, indeed, one would have to return to a subject that one had thus left
behind:

'. . . technology cycles too doesn't it? A thing which is impossible now,
which you have done some basic work on – you put on the shelf, and it comes
back in three or five years' time, so you have to consult the early experts
again'.

This was regarded as quite difficult, because of scientific progress in the interim
period – as one scientist put it: 'You simply spent a large amount of man-hours
simply re-inventing the wheel'. Nevertheless, a reputation for being a specialist
on a particular topic could never quite be lost:

'I agree with – over this business of your old job never really leaving. Only
yesterday I had an example of that. Someone came in and said "How do you
make [Xs]?" and I said "Well, excuse me, but why have you come to
me?" . . . She said, "Well, you had something to do with [Xs]", and I said
"Yes, but that was fourteen years ago". I've never looked at an [X] since.'

But the typology of 'problem choice' in Table 1.3 gives a misleading impression
of precision. As we saw in §1.2, the cognitive 'map' of a scientific discipline is
a very inadequate guide to the actual relationships between specialized research
topics at any given time. The categories under which scientific papers are
classified for information retrieval are often quite arbitrary, and suggest
boundaries that would not be evident in reality. In any case, the formal definition
of a particular 'problem area' in the official classification scheme of an established
science such as astronomy may not be revised for 20 or 30 years. In that period,
the problems covered by that definition and the methods used to solve them may
well have changed out of all recognition. Thus, even the scientist who continues
to publish papers in that area throughout his or her research career has probably
had to make radical changes of theoretical approach and technique in the course
of those years.It is the duty of the acknowledged 'specialist' on any subject to keep
up to date: if that subject moves, so to speak, over the scientific landscape, then
the specialist must move along with it, and acquire new knowledge, new skills,
and new scientific interests along the way. A physiologist made this point very
clearly:

'The problem has not changed. From my PhD onwards it's been the same
problem, mainly because I was very much influenced by one person as an
undergraduate, I think . . . . I have been focussing on one particular aspect
of it – the relationship between the brain and the spinal chord. But the tech

niques and approaches have changed considerably in the time I have been doing it . . . .'

Similarly, a specialist on one aspect of a major scientific technology described a long career, with transitions from one problem to another, and from one R&D establishment to another, all essentially in the same narrowly technical 'specialty'. Another scientist in the same technology had to be transferred organizationally from physics to chemistry as he followed his research on the properties of a particular material of great importance in that technology, thus apparently moving a vast distance on the disciplinary map without really changing his specialty at all. In practice, there is little distinction between this process of following a 'subject' as it moves across the scientific landscape and the complementary process of moving along a chain of adjoining or overlapping specialties. Again and again, scientists described the evolution of their research interests as an unplanned process of gradual *drift*:

' . . . it started off being very specialized . . . . I had to work on very specific high pressure − measurements. Then after that I became interested in − solutions in general. So that broadened it out . . . , and then I became interested in . . . why the waters have the composition they do. And so I am doing quite a range of [research] now which isn't directly related to the work I did in my first job . . . . The techniques have changed a lot. I am not working at high pressures any more, and . . . looking at a wide range of chemical techniques, rather than just potentiometry . . . I have become interested in more complex systems.'

'I have changed gradually over the years . . . . I started off as a botanist . . . . When I had a permanent post it was to do with vegetation classification . . . . But that gradually diverged from applying the quantitative methods to a much wider scale . . . . And so where I am now is much more concerned with looking at ecological impacts, and seeing how ecology can − well, the relationship of ecology to modern society . . . . It's a long way from the early days of botany. I am still just as interested in botany. In my holiday I still potter around looking at plants.'

The same experience is echoed round the group:

'[I did my PhD] on the physiology of vision − how photo-receptive cells in our heads respond to photons, or whether they can electrically detect single photons. That's what I started to do. This has sort of gradually evolved into an interest in nervous system development, and I just sort of adopted, like everybody else, new methods.'

Q. 'Would you say that what you are doing now is quite different, quite varied?'

'Utterly different from what I started off with, and I'd be rather sad if that weren't so. I don't think I would like to be drafted back into that . . . . What I mean is, you are a physiologist. What I do is make up solutions and bottle wash, where beforehand I fiddled with electronics, and I think that is quite different. I think the range of skills and knowledge are utterly of a dissimilar kind.'

[Another speaker] 'Yes, that is a very important point, I think. The titles people bear are always totally irrelevant, very often, to the thing that they do. I mean, I am a microbiologist. I did this for a lot of my career. What I am doing now has no relation whatever to I started doing, partly because of the inherent change in the [subject] itself.'

Indeed, the drifting trail of many research careers cannot be charted properly on a conventional 'map' of science. Each 'field' or 'subfield' adjoins or overlaps many others in a manner that cannot be represented correctly on a classification tree, or on a two-dimensional diagram. Take the case of the agricultural researcher whose specialty was the study of the natural defences of a particular type of crop to a particular group of diseases carried by a particular group of insects. For various reasons, he 'started to get interested' in the same phenomenon in a different type of crop, so 'the change really was getting to know a crop rather than changing my research interest'. But he might equally have changed to a slightly different group of diseases, or to a different type of insect vector, or begun to look in more detail at certain physiological or biochemical mechanisms, and thus gradually drifted off along quite a different trail. The multiplicity of the connections between scientific specialties will turn out to be a key factor when scientists are forced to enter unfamiliar areas of research (§4.4, §8.2).

1.8 Scientists as life-long specialists

The division of scientific labour into innumerable subject specialties gives rise to a naive conception of scientists as people who devote their whole lives to the study of some very obscure and very detailed question, until, as the old joke has it, they 'know everything about nothing'. As we have seen in this chapter, there is an element of truth in this stereotype. Given the freedom to do so, the natural tendency of most scientists is to concentrate for years on a few problems in a narrow area of research. Not only does such 'persistence' seem to them the only way to make progress in the pursuit of knowledge: as we shall see in later chapters, it is usually strongly reinforced by career considerations, such as the desire to gain reputation, recognition and promotion (§5.3).

Yet they do not seem to revel in this situation, and often show their uneasy awareness of the risk of becoming totally committed to a highly specialized activity that is bound, eventually, to come to an end. The serious personal consequences of *undue* persistence will be discussed in later chapters (§5.6). Even scientists who are quite free to make their own research plans have to adapt to the ever-changing landscape of knowledge itself. The delicate balance between specialization and diversification is brought out in the following discussion:

'If you are in mid-stream, say, along a large project in a certain area, you are not receptive necessarily to the outside possibilities because you are occupied in what you are doing. When you are coming towards the end of a piece of work and you are finishing it off, you may rather open up your mind to other opportunities. And when these opportunities arise, that's the time you will be receptive to them. So it's not a question of people being receptive all the time. If people are receptive all the time they never seem to get anything done in any field, because they keep hopping around too much.'

[Another speaker]. 'Well, I don't think that's true. I think you can be receptive without doing anything about it. I think you can be aware of the possibilities all the time, and know that you've got to soldier on, and get the piece of work that you are doing done. But I think you can still keep your nose to the ground and be aware of what's happening in the organization, and what's likely to happen, and make a few intelligent guesses.'

This 'receptiveness' may also be fostered by the knowledge that they must always keep their minds open to new questions that are relevant to their particular scientific interests. Even the narrowest and most precise scientific problem or technique has connections that proliferate widely across the scientific 'map'. The remarks of an ecologist bring out this point:

. . . the intellectual exercise of trying to create something which is a parallel of something in the real world, I find a big challenge and of great interest, and this has really kept me going through a number of research projects which on the face of it are very disparate, and have very different organisms involved in different habitats. But really they can be grouped together under this heading . . . I think we gain experience for different habitats as ecologists. You can get a feel for particular habitats and I think also a love of particular habitats and that lasts all the way though. My PhD research was on the [--] habitat, and I sort of gained a feeling for that which I tried to keep going. Whenever I have a chance to . . . get involved in that particular habitat I take it. Having said that, if somebody else comes to me with a problem involving a different habitat, or I see an opportunity for some support to work in a different habitat . . . I am still happy to go into a different habitat. I am not tied to being a [--] person, but I still retain this love of [--], so there is an evidence of conservatism. But at the same time, you

know, one is prepared to branch out a bit if somebody comes to me with
something outside . . . my own field . . . I think that is the same
philosophy'.

Not all scientists are narrow specialists. As we have seen, there are an appreciable
number who seem to spread their research interests over quite a large area.
Everybody knows the alternative stereotype of the scientist as a person of in-
satiable curiosity, always ready to drop what they are doing in order to find the
answer to any scientific question that happens to occur to them. Such scientists
do exist, but this stereotype, also, is something of a caricature. As the above
quotation shows, a research trail that looks quite diversified to an outside observer
may actually have been generated in pursuit of a single conceptual theme or in the
exercise of a single experimental technique.

The large factor of chance in all scientific work is evident in all the accounts
scientists give of their careers. Yet they emphasize the element of deliberation
in the way that they deal with this uncertainty. Looked at from the outside, a
research trail may appear quite erratic; yet for the person who actually followed
it, it may have had a piecewise rationale:

'I think the ways in which I would like the research to go have always been
fairly obvious, in that I haven't come across any obvious cross roads
wondering which way to go: it's always been quite clear which way to go.
It has not always been in a way that I thought I would be going three months
ago, [but] I have never come across a point where I wonder which way to
go next . . . In my own experience, it's been, really – the choice seems to
have been made already, in a sense.'

Not every scientist would feel that such decisions were really as deliberate as they
might look in retrospect, but most would agree that they usually have plenty of
questions on which further research might be done if the occasion demanded. As
somebody put it, evidently referring to one of the characteristic dilemmas of 'problem choice' in science:

'But it is also true, we are in a field where there is no sudden dead end. I mean,
normally, research evolves, you know, in various directions, and you can,
you know, quickly, if you see a dead end at one end, you jump to another
branch; the whole thing is progressing in the same direction. I don't think
we are in a field where you can identify a sudden end – a sudden block to
all possible routes. But I would have thought most of the programmes were,
in fact, constantly changing, constantly evolving.'

It is not to our purpose in the present work to analyse in detail all the considerations that enter into the answers that scientists give to the question they must
continually ask themselves: 'What research shall I do *now*?' (Ziman 1981). But
the point that comes out in all their accounts of their own careers is the great

importance they attach to an ideal of intellectual *continuity*, coupled with a sincere appreciation of *gradual* change in the subject matter of their research:

Q(?).'So it's a gradual change . . . ? At point X you are doing one thing, and at point Y you are doing something completely different. In between there are a lot of small stages where you gradually move from one to the other.'

B(?). 'Yes, I think that is usually the way. I don't really think it is catastrophic at any time.'

In another interview, somebody said:

'These changes have got to be made in short stabs. After all, my career — if you think back to where I started, in coal mining, to where I am now the contrast is considerable, but at no stage along that line has any step been very great. I have been able to take small steps at a time, and it's just evolution really.'

It is this sense of the gradualness of *natural* change in a scientific life that is at the heart of our present investigation.

2
The research system

2.1 The collectivization of science

In the previous chapter, scientists were deliberately portrayed as *individuals* working independently on narrowly specialized research problems distributed over the broad 'landscape' of science. This naive image was presented in order to bring out a number of features of the scientific life that are relevant to our present enquiry. But scientists actually work together in a variety of *establishments*, where they are employed as professional researchers, technical experts, administrators, managers and teachers. These establishments (or 'laboratories', 'institutes', 'research stations', etc.) are answerable, in turn, to larger *organizations* such as industrial firms, government departments, research councils, and universities. They are clearly so potent that they cannot be treated simply as passive frames around the scientific life: their practices and policies decisively shape the careers of most of the scientists whom they employ.

This is a relatively new development. Until, say, the Second World War, the majority of scientific research was carried out in the traditional 'academic' style, where each researcher was free − at least in principle − to undertake any investigation that he or she thought worth while. Scientific work has always been so highly specialized that most scientists have usually followed narrow research trails, but this was by their own choice. In practice, many social considerations might have to be taken into account in a decision on what research problem to tackle next (Ziman 1981), but such decisions were seldom determined by external agencies. The scientific community was, in fact, more authoritarian, and more tightly organized into 'invisible colleges' and 'specialties' than might have been suggested by its official spokesmen (Ziman 1984), but it was not systematically organized and had no general plan for the research to be undertaken by its members. The direction in which science happened to evolve was very largely the outcome of innumerable personal decisions by individual scientists, competing for jobs and for esteem.

One has only to look into any modern research laboratory, and listen to scientists talking amongst themselves, to realize that there has been a profound transformation in the way that scientific work is now organized. Internal and external forces have combined to 'collectivize' the research process (Ziman 1983, 1984). Gigantic instruments, such as particle accelerators, and immense technological projects, such as the development of satellite communication systems, demand the efforts of large teams of research workers (§7.4), with a wide range of specialized skills. In this sort of 'big science', there is usually such a mixture of scientific 'research' and technological 'development' that one can only speak of a general activity of 'R&D'. This applies even to 'little science' subjects, such as microbiology, where technological opportunities are continually being perceived, and where elaborate administrative arrangements have to be made to provide small groups of researchers with all the sophisticated apparatus that they now need. This very expensive activity is supported by governments and commercial firms primarily for the practical benefits it can offer for industry, defence, agriculture, medicine, etc. and is therefore programmed, planned, and controlled to achieve those benefits.

This change in both the internal structure of science and its external social relations is obviously an extremely significant historical process, calling for detailed sociological, economic and political study (EASST 1984). Indeed, one of the purposes of this book is to observe and analyse just one aspect of this change − its effects on the personal careers of the people involved. To make any progress in this particular investigation, we must therefore regard the overall process of collectivization as a background of change whose rationale or consequences are not, in other respects, to be questioned. Whether for good or ill, the fact is that, in Britain as in most other advanced industrial countries, scientists now work mainly in large 'R&D organizations' or 'technical systems' (Shrum 1984) whose goals are set either by non-scientific bodies such as government departments and boards of directors of companies, or by high-level scientific bodies such as research councils. To set the scene for the actors in our drama − the scientists themselves − we need to look briefly at some of the general characteristics of these organizations.

2.2   R&D organizations and projects

R&D organizations come in all sorts of shapes and sizes. The British Ministry of Defence is a rigid bureaucracy, employing something like 10,000 graduate scientists and engineers: a quasi-autonomous research group in a university may consist of a dozen or so, more or less independent researchers, loosely guided by a senior academic. It is obviously very difficult to make general statements about such a wide range of organizations.

Nevertheless, from the point of view of the individual scientific worker, they have a great many features in common. They all offer careers to people with very high educational qualifications – very often PhDs – in scientific and technological disciplines. The work to be done is of the general nature of 'research', in that its primary purpose is to produce new and reliable *knowledge* – whether for immediate use or 'for its own sake' – rather than to manufacture material objects, to provide routine services, or to manage the work of other people. Whatever its purpose, and wherever it is carried out, this work is based upon the same general scientific principles, draws upon the same archives of fact and theory, uses the same techniques and instruments, and has to satisfy the same criteria of technical rationality.

These common features of all scientific work are what makes sense of the notion of a specifically *scientific* career. In fact, scientists are more sharply differentiated by their subject specialties than they are by the type of R&D organizations in which they work. The lecturer in botany who could not conceivably get a job in the university astronomical observatory across the street might be considered an ideal candidate for a senior post in one of the research establishments of the Department of the Environment, or in the research laboratory of a company manufacturing pesticides, perhaps on the other side of the world. At the level of day-to-day or week-to-week activity, there might be practically no difference in the work to be done, at the laboratory bench, in the library, in consultation with colleagues, or in the privacy of one's own thoughts.

This was confirmed on many occasions in the interviews. A scientist in a government R&D establishment made this clear:

'The job I was doing at university was very similar to that which I ended up doing here. The equipment and facilities are very similar in terms of workshop support and computing support. In fact, computing support is far superior to university. And the sort of in-house facilities are very easy to get at.'

The same point was made at another, semi-industrial establishment:

. . . the hours are a little different: that takes a little bit of getting used to . . . . But I think working conditions are very much the same, and the facilities are made available, library facilities, very much the same as in universities – possibly a bit better'.

Should 'laboratory life' be treated, then, as a uniform activity, much the same wherever scientists work? This is apparently taken for granted by recent ethnographic studies (Latour & Woolgar 1979; Knorr–Cetina 1981). These studies have been made within such limited observational frames of time, space, and social interaction that they can do little more than draw attention to the sheer ordinariness of scientific work, and have not been sensitive to possible differences in the social atmosphere in different R&D organizations, or even between different laboratories in the same organization. Comparative studies of managerial practices (Lemaine, *et al.* 1972: Shinn 1980) have shown, however, that large differences of this kind certainly exist, and this is fully supported by the words of scientists themselves, as will be apparent from many of our excerpts.

Some of these differences are undoubtedly systematic. R&D organizations in Britain fall fairly neatly into four 'sectors', according to the way in which they are controlled and financed (see §2.3), and this has a large effect on their internal managerial structure and personnel policies. But the difference between, say, a career as a scientist in a university, and a career as a scientist in a big company, is not entirely due to a difference in the nature of the job to be done: it may arise in large part from the fact that the 'academic' and 'industrial' sectors of society follow very different managerial and administrative traditions. This, again, is a point that will be taken up later in more detail (Chapter 6).

What most people would say is that the way in which the scientific work is managed in a particular R&D organization should be simply a function of the overall objectives of that organization. If this objective is to produce 'pure' scientific knowledge, just 'for its own sake', then the work should be managed 'academically': if the objective is the more practical one of 'applying' science for profit or other immediate benefits, then an 'industrial' style of management is appropriate. There is much to be said for this point of view, but the implication of a sharp line of demarcation between the 'pure' and 'applied' modes of scientific work is simply not consistent with contemporary realities. One of the main characteristics of 'collectivized' science is that the objectives of R&D organizations do not fall neatly on one side or another of this divide, and can only be placed very vaguely along a continuous spectrum ranging from very basic research, of no foreseeable utility, at one end, to frantic development work, to meet immediate commercial or political needs, at the other (Ziman 1984).

But even a broad classification of R&D organizations in terms of the social 'relevance' of their objectives does not bring out the features that really characterize them as places in which to do scientific work. Whatever may be said about the objectives of a large 'technical system' in its official charter, the actual work that has to be done to achieve those objectives may have a very wide range of immediacy or apparent utility. It turns out, moreover, that scientists are not much concerned about whether or not the particular problems they happen to be studying are 'relevant' in this sense (Jagtenberg 1983). As their conversation continually shows, their main concern is with the technical aspects of their jobs, and whether they can carry out in practice the tasks that they are expected to perform. In other words, they tend to describe the organizational framework of their work as if it were simply the source of the specific R&D *projects* in which, more or less voluntarily, they happen to be involved.

This notion of a 'project' is not well-defined, but it suggests a coherent unit of scientific work that can be prescribed in advance and eventually completed. Such a prescription might be to find the answer to a particular scientific question or to solve a particular technological problem. R&D projects clearly differ considerably in the extent to which they are precisely specified. But even in 'academic' research, which is supposed to proceed without any plans or programmes, grants are awarded on the basis of detailed project proposals which specify the problem to be tackled and the likely significance of the results (Ziman 1981). The way in which scientists nowadays refer to 'programmes' and 'projects' clearly indicates the general terms on which scientific work is normally broken down into manageable units and presented to them within R&D organizations.

2.3 Urgency and extent

R&D projects obviously vary greatly in scale and intensity. It is convenient to represent these variations in terms of two parameters — 'urgency' and 'extent'. These parameters are strictly notional, but if one were to attempt to define them quantitatively one might say that the *urgency* of a research or development project is in inverse relation to the time needed to complete it, whilst its *extent* would be proportional to the number of people directly involved.

The concept of urgency can obviously be related to the degree of 'relevance' attached to the work. For example, 'pure' research in the traditional academic mode is simply not urgent at all; if the questions under investigation are being studied solely 'for their own sake', then no date at all need be set for answering them. The modern concept of 'basic' research is similarly timeless in principle, but when research is described as 'strategic' it is expected to yield useful results within the foreseeable future — a few decades, say. In 'mission-oriented' research this time scale is shortened to a few years, whilst in technological development the solution to an urgent problem may be needed in just a few months.

But although 'urgency' is often roughly proportional to 'relevance', it is more specific, since it can be applied operationally to the work in hand and does not have to be referred to vaguer and more distant objectives. Thus, for example, the design of a novel cosmic ray detector for a space probe would count as a very urgent project, since it would probably have to be undertaken to a very tight timetable, even though the enterprise as a whole was not expected to produce any results of practical human value. Conversely, even in the research laboratory of a big chemical company, where all the work is relevant to the development of new products and processes, a few scientists would be engaged in exploratory research with no time constraints and no specific objectives.

The relative urgency of the research projects in which he or she is involved must obviously play a very important part in a scientist's career. The rate at which it may be desirable, or even practicable, to change one's problem area must depend on the timescale for the solution of problems within that area. If it is going to take 20 years to make real progress on a particular type of problem, then it would be professionally irresponsible to 'migrate' to another field (§1.6) after only five or ten years of research. On the other hand, a scientist working in an organization where most of the projects are very urgent is forced to 'diversify' his or her research interests (§1.5) in order to maintain a steady flow of work.

In this connection, it is essential to distinguish between intrinsic and extrinsic conceptions of 'urgency'. If scientific research were largely the routine application of standard procedures — like surgery, say, or banking — the time required to solve a well-posed problem could be estimated from experience. In fact, one of the tacit skills of scientific specialists is the ability to estimate how long it will take to answer typical questions within their fields. When a scientist says, for example:

> There are artificial rules for us. We get research programmes that are supposed to last three years.'

He is pointing to a mismatch between his notion of the natural rhythm of his research (Hargens 1975) and the patterns imposed on it for administrative purposes. But this concept of an 'intrinsic' urgency is not easy to articulate, and is often overridden by the 'extrinsic' urgency imposed by the organizational milieu, where the time-span allotted to each project is strongly influenced by other factors, such as the availability of resources and people. Not surprisingly, the urgency of projects often becomes a matter for negotiation between researchers and their managers, since it essentially determines the pace and intensity of the work to be done.

One of the most obvious implications of the 'collectivization' of science is that scientific work is done less by individuals than by teams. The 'extent' of a project is simply the size of the group of researchers working together on it. This is obviously a fundamental parameter from a career point of view, because it determines the range of skills required of each member of the group, and the degree of autonomy and personal discretion permitted in the exercise of these skills. The size of the group also affects the skills required of those who aspire to lead it, with the prospect, no doubt, of moving out of 'bench research' into higher managerial roles (§6.2).

Here again, it is essential to distinguish between the actual extent of an R&D project and the extent of the administrative subdivision of the R&D organization where this project is being undertaken. In some cases members of a 'section', or 'unit' are not really working closely together, but are involved in a variety of small projects in an individual capacity. In other cases, an *ad hoc* team drawn from a number of administrative subdivisions may collaborate for a long period on a major investigation with only a loose managerial structure to hold them together (§7.4). Thus, the actual extent of the projects in which a scientist is engaged is a significant feature of work in an R&D organization, but it cannot always be determined from a superficial scrutiny of the administrative chart of that organization.

'Urgency' and 'extent' are not, of course, independent variables. From a practical point of view, the extent of the team assigned to a project usually depends on the urgency with which a result is desired. The general policy in basic research, for example, is to keep such teams as small as possible, given the amount of work to be done and the range of specialized expertise required. On the other hand, the technological development of a new industrial product is often so urgent in its final stages that very large teams are assembled to push the work through. Conversely, the extent of a project may impose a considerable degree of urgency on it. Thus, in some fields of 'big science', such as experimental high energy physics, the projects are not intrinsically urgent, but they are so large that they can only be undertaken by a correspondingly extensive team. To coordinate the efforts of such a team, each stage of the project has to be carefully programmed, and made urgent for those working on it.

This *segmentation* of R&D is characteristic of all very large technological undertakings, such as the design, development and demonstration of aircraft, power stations, chemical plants, weapons systems, etc. Such undertakings often involve the labour of thousands of scientific specialists, over a period of a decade or more. This labour has to be divided, often long in advance, into numerous specific projects, each of the appropriate urgency and extent. In the development of a nuclear power plant, for example, research on a new canning material for the fuel elements is set certain technical objectives, which must have been reached before certain problems concerning the disposal of burnt-up elements can be tackled, and so on. For researchers working in such organizations, the progress of the undertaking as a whole is largely irrelevant to the way in which their own work is driven along, and coordinated with the work of others.

2.4   Sectors of R&D activity

R&D organizations in Britain are not integrated into a national 'R&D System' with standard employment practices and an overarching plan. For statistical purposes they are usually classified according to the politico – economic *sector* to which they belong. This classification is not to be taken as definitive, since it may hide important differences between organizations within the same sector (e.g. between the research divisions of different industrial firms), and even between different establishments belonging to the same organization (e.g. the various R&D Establishments of the Ministry of Defence). Nevertheless, each of the four main sectors has its characteristic objectives and administrative arrangements for R&D.

Universities and polytechnics

By tradition, the permanent academic staff of universities and polytechnics are expected to undertake a certain amount of research in parallel with their work as teachers. This expectation is not a formal condition of employment, but is well understood to be the major factor in earning promotion and other personal benefits (§6.5). It is thus mainly undertaken on an individual basis, and need not be specifically directed towards the solution of any practical problem. In the past, these objectives could usually be met by doing long-term, basic or strategic research in small groups. In other words, 'academia' is characterized by relatively non-urgent projects, of limited extent. Not every academic is as conscientious as the one who said:

'I at times certainly have the feeling of being rebuked, with the freedom I have to do really what I want to do. In that sense society has really been very good to me. And I think, in recompense, what I intend to do is to devote part of my time to actually [tackling] immediate problems which I think the society would benefit from, but leave enough of my time to explore the areas which are of interest to me, and this is my kind of trade-off. I do this personally, and I am aware of it.'

The managerial tradition in academia is nominally a nullity: each academic scientist is supposed to have complete autonomy in the choice of research areas and research problems. This is surely a myth, which would not be confirmed by detailed study of the relationships between university professors and their junior colleagues (§6.3). Nor does it apply to the many scientists now working in academic institutions as full-time researchers, on temporary or permanent appointments.

In any case, this traditional individualism is now firmly bounded by the practical need to get financial support for research from outside bodies such as the research councils, government departments, or industry. The standard procedure for obtaining a grant for research is to propose to answer a specific scientific question within a specific time − in a word, a 'project'. But there is fierce competition for funds for research, and not every project of every academic scientist can be sure to win support. That is to say, scientists in academia are forced to harmonize their research plans with the objectives of funding organizations, even if they are not directly 'managed' in their day to day work.

Research Councils

These are organizations charged with advancing broad fields of basic or strategic research in relation to Agriculture and Food (AFRC), Medicine (MRC), the Natural Environment (NERC), Science and Engineering (SERC), and Economic and Social Studies (ESRC). Although these Councils are not, strictly speaking, governmental bodies, their members are appointed by the Government, and they derive their funds primarily from the Treasury. One of their important functions is to distribute a large proportion of these funds as research grants to scientists in academia, as indicated above. But they also run their own research establishments, where a large number of scientists are employed full time on research. The appointments of the staff of the AFRC, NERC, SERC and ESRC parallel those in the Scientific Civil Service (§3.3); staff grading in the MRC parallels university grades in medical disciplines.

The original purpose of the Research Councils was to complement or supplement university research in disciplines, fields, or types of science which were thought to show a high degree of 'timeliness and promise', or where there was seen to be a strategic national need for scientific knowledge. This could include entirely 'non-urgent' projects in 'pure' science — for example, the research in high energy physics and astronomy undertaken or funded by the SERC — but, as their titles indicate, the various research councils are expected to carry out research that can be related to practical social needs.

Until recently, this criterion of 'relevance' was not translated into a high degree of 'urgency'. Research plans, in general, came from within each establishment — often, indeed, from individual scientists or heads of small research groups — subject to broad approval or orientation at Council or Board level. Research projects were not necessarily very 'extensive' in terms of staff, but they often lasted many years, with periodic review for resource allocation in competition with new projects.

A research council scientist may sometimes feel that:

. . . there is a certain attractiveness about a university, a nice cushy life, in some cases. Particularly the Oxford environment. One could go and have nice evening meals in college, and that sort of thing, which one misses here. No common room here.'

But this nostalgia did not apply to their scientific work. A scientist at a highly esteemed research council establishment was frank about it:

'When I talk to colleagues in university departments [in my discipline] now I find them very envious of our being able to do full-time research. They can manage it and fit in the teaching, and so on, but generally we seem, in our particular line of country, to have better facilities, more time — obviously more time — and so on, and that's the general conclusion. They always seem to be saying: "You lucky chaps here; you can do this full time, and we can't, and that of course makes it less hard to produce these papers which seem to be so important".'

Until the early 1970s, scientific life in this sector could thus have been described as 'quasi-academic' (Ziman 1981). But as a result of the Rothschild Report (§2.5), three of the Research Councils (ARC, MRC and NERC, as they then were) were forced into a new financial regime, whereby a substantial proportion of their funding for 'in-house' research had to be earned through specific 'contracts' from 'customers' — mainly the Government Departments for Agriculture, Health, etc. As a result, the balance of scientific effort in this sector has shifted from long-term 'strategic' projects to much shorter-term 'commissioned' research (§2.5). As we shall see, this is only one manifestation of a general trend, in all sectors of R&D activity, towards greater 'urgency', which is putting considerable strain on individuals and institutions.

The public sector

A considerable proportion of the research scientists in the UK are directly employed by the central government. The *Scientific Civil Service* is spread over a number of Ministries, the largest groups being in Defence (10,700 staff in scientific grades), Agriculture and Fisheries (1900), Industry (1800) and Environment (1200). Its members are normal civil servants, employed under standardized conditions of pay and promotion, although they form a recognizable group with appropriately labelled grades, such as Scientific Officer, Principal Scientific Officer, etc.

Generally speaking, government R&D work is concentrated in a small number of relatively large establishments, each managed and administered on conventional bureaucratic lines and directly responsible to Whitehall for financial resources, research plans, personnel policies, etc. This is characteristic also of the research establishments of the UK Atomic Energy Authority, which are not, of course, technically part of the scientific civil service, but which have similar objectives and functions in relation to the nuclear power industry and are therefore most simply classified in this sector.

R&D in the public sector is extremely varied in its objectives. It can range from 'strategic' research of very general relevance, such as climatology, to the solution of major technological problems, such as getting a new air-defence system into operation. But it is not supposed to be involved in the search for knowledge 'for its own sake', nor in the final stages of industrial innovation. In some establishments a considerable amount of the scientific work is to provide routine services to industry and agriculture, local government, etc. (§6.4). As a consequence, there may be a general background of numerous small research projects and routine technical tasks, enlivened by a few major projects that are much more 'extensive' and 'urgent' than would be normal in academia or in a research council establishment. Because of their scale and complexity, such projects may actually take many years to complete, and they are continually subject to re-appraisal, and may be started or stopped, somewhat arbitrarily, at relatively short notice. Nevertheless, the traditional atmosphere of R&D in the public sector is quite distinctive:

'I think this Establishment in particular has a very good reputation for picking up bright ideas and so on.'

'And the freedom of action to be able to carry them out. Provided . . . they make a case to the management, then there is a high degree of freedom to work on various things − much more so, I think, than in industry. Indeed, I know one or two people in industry who actually left to join the industry because they felt that they are directed.' . . .

. . . 'You can ask: "Why are we civil servants: why aren't we in industry or universities?". I think perhaps it's because somewhere inside all of us we prefer to be secure, rather than moving around as they do in industry.'

'I feel less secure these days than we did.'

'Oh, certainly. But still, compared with industry I think we are relatively secure. Why didn't we go into industry, any of us? We certainly get more freedom here.'

'. . . I mean industry doesn't really do research: it does short-term development.'

Notice that the comparison is always with work in industry. Twenty years ago, scientists working in this establishment would probably have thought of university research as the obvious alternative to the scientific civil service.

The private sector

The R&D divisions of a number of relatively large industrial firms (including, nowadays, the nationalized and denationalized industries such as British Steel and British Telecom) constitute a dispersed and diverse environment for scientific employment. For example, there is nothing like the standard hierarchy of 'scientific officer' grades, as in the Civil Service and the Research Councils (§3.4). Each firm has its own personnel policies, its own system of appointments and promotions, its own managerial structure.

But the nature of the work itself is much the same, across the whole sector. Primarily, it is technological development, with a backing of 'mission-oriented' research directed towards product or process innovation. Basic research, or long-term strategic research is regarded as a luxury, which only a few very large firms can really afford or profit from. Most innovative projects are 'urgent', and often very 'extensive', so that team work under firm management is the norm.

An atmosphere of urgency may also be generated by the day-to-day demands of the production divisions of the firm, which may turn to the R&D division for help in 'problem-shooting'. In fact, a large industrial firm may employ scientists in a variety of jobs, ranging from fairly fundamental research, through technological development, to technical roles in production and marketing.

In the private sector, projects must be proposed in relation to their productive outcome, and assessed according to their potential profitability to the firm. As a consequence, the programme of the R&D division of each firm is subject to sharp and sudden shocks, deriving from progress in the 'state of the art', the commercial fortunes of the firm, general economic conditions, national industrial and fiscal policies, etc. In this sector, therefore, changes of project, of job assignment, of research objectives or of product specification have always been regarded as normal.

2.5   Stresses and strains in the R&D system

Change is of the essence of science. An R&D organization that is not under some pressure to change is probably not doing its job, and should be closed down at once. These stresses have increased markedly in the last 20 or 30 years. External forces press more and more heavily on the science they have 'collectivized' (§2.1). On the one hand, society is ever more conscious of new problems that science might solve and new capabilities that it might achieve. New research programmes are rapidly drawn up, and instructions given to R&D establishments to proceed with them as fast as possible. On the other hand, vast technological projects are conceived by industries and governments, and then suddenly cancelled, after years of R&D effort, in response to economic and political circumstances that were not foreseen when they began. The recent history of many R&D organizations in the private and government sectors is punctuated by the effects of such decisions, over which they have had practically no control.

These stresses also arise from within R&D organizations, as they perceive new opportunities or become disenchanted with older projects. Research itself is a progressive enterprise (§1.2), in which one can never be allowed to rest on one's oars. In a very active field, the solution to an old problem simply becomes the basis for an attack on a new one. A breakthrough generates a tremendous urge for an increase of effort:

' . . . at the present moment . . . it's just expanding so much, there are so many different fields which we would like to play around with without limitations of space and manpower, and the problems are so incredibly complex, that . . . as long as we can continue with funding and enough people to do it − you know . . . within my lifetime there is no way we are going to come to a straightforward routine piece of work'.

Similar pressures can arise when the time seems ripe to exploit a 'bright idea'. People want to get in on it before the fashion spreads and it becomes a 'bandwagon'.

On the other hand, there is often a feeling that it would be wise to terminate a particular project or research programme that is running into insoluble difficulties, or is becoming scientifically uninteresting, or is not likely to win support from the higher echelons of the organization, and to move to some other problem or problem area:

'I think that was what [the previous director of the establishment] saw very clearly − that the opportunity of getting funds [for] just monitoring [an environmental hazard] would run out and therefore it was time to build a viable alternative.'

The Medical Research Council has a long-standing policy of drastically reviewing the programme of each of its research units when the director retires, with the initial presumption that the unit should be closed down in order to free resources for new research initiatives. In some R&D organizations scientists themselves complain that their own plans for change are resisted by external authorities:

' . . . it's very difficult to get a new project off the ground, but it's much, much easier to get an existing project extended. So it's much easier for us to think of adjuncts to existing projects which we shall almost certainly get funds for. But if we try to to dream up new directions the Department and the funding agencies, whoever they may be, tend not to like that change in direction'.

It is quite wrong to suppose that scientific life is peaceful and unchanging, and that scientists dislike to be disturbed in their research.

In the last few years, the stresses of normal scientific and technological change have been amplified by financial retrenchment and administrative reform. The material and human resources devoted to science, which had been growing steadily and rapidly ever since the War, have now levelled off, or declined. In every sector of R&D activity, from universities and polytechnics to government and industrial laboratories, research budgets have been drastically cut and scientific jobs lost. Projects have to be carried on, or started, with inadequate resources. Even when an establishment is granted the funds for a new project, it may not be permitted to appoint new staff to undertake it. In some R&D organizations, these pressures are now so intense that active research units − even whole establishments – are being closed down, or 'privatized', or instructed to shed staff, or moved unceremoniously to quite different lines of work.

These financial pressures have also produced a significant change in the general climate of research. Funding authorities – government departments, company directorates, research councils, etc. – scrutinize the budgets and programmes of R&D organizations much more closely than they used to, trying to save those projects offering the most tangible benefits at the expense of those whose outcome is more doubtful. In other words, the whole R&D effort of the country has been shifted towards projects of greater 'urgency' (§2.3). In the private sector,

'. . . this is the case in all industries. All the industrial companies that I come into have similar problems to the ones that we have. They are cutting back on their R&D, they are cutting back proportionately [sic] on the fundamental end of R&D, and in many companies it's disappearing altogether. In the larger companies it is being cut back'.

The same is happening in the government sector, where, for example, the Ministry of Defence has reduced its expenditure on basic research to very low levels (*Reference?). In their turn, the research councils are putting into practice the policies advocated by a very senior scientist to a member of an ARC establishment:

[His view is that] in agriculture we've got enough basic science at the moment; what is really needed is that this should be applied. That is what he said – and he said it loudly on many occasions – that if we are going to improve productivity in the UK we should be applying what information we already have: that's how we should do it. And he looked around and he saw so much inefficient farming, and said, ''We know the science, let's get this science through to the farmer and get them educated . . . ''.'

Even in the academic sector, where research is not centrally directed, university scientists have been encouraged to make up for shortages in the funds that used to come from the UGC and the Research Councils by taking on R&D work for industry.

This is not the place to discuss the long-term consequences, for science or for the nation, of this decisive shift towards more 'urgent' and 'relevant' scientific work in Britain (Irvine & Martin 1984). Our concern is simply with its effects on scientific life, and on scientific careers. Some of these effects are quite direct. As an industrial scientist remarked:

'Long-term planning does not exist as such. We are working on a very short time scale, and that affects contact [with scientists in universities]. How can you possibly make contacts [when] you've got, say, six months, twelve months, to solve a problem . . . ?'

Projects are driven along much faster than the scientists themselves believe to be reasonable (cf§2.3):

'We can even say now what we ought to be doing, and it's all the more frustrating that we have a time scale imposed on us by [the relevant Government Department] which completely destroys the natural evolution of the research that you [*sic*] have in mind. As far as I can see, the research has a certain pace, which goes out naturally because of the people involved, and the amount of money, or the . . . manpower available for it. That's what is screwing us up at the moment. We, on the one hand, have a pace we want to go at, and, on the other hand, we are not allowed to do it.'

This is very apparent in the academic sector, where institutional resources for research are now very limited, and most scientists cannot undertake serious research at all unless they can get a grant for it from a research council, government department, charitable foundation or industrial firm. Overall reductions in the funds allocated to basic research means that the competition for such grants has become very intense, and the procedures for the selection of projects by 'peer review' are severely strained. The money that is available is spread too thinly, with predictable consequences:

'[I think there is a] very real problem of continuity in research in universities because of funding. A lot of our research is done with the aid of research assistants, or research students to some extent, and this is very much a three-year business. I have been very frustrated in the past, in the sense that techniques, methods, and so on, have been developed, results are beginning to come through, and I have had trouble in getting further funds to continue the programme. Everything has come to a halt for a while: then, two or three years later, I have managed to get a further grant — but many of these techniques have had to be learnt again.'

In the public sector, the move towards greater urgency is driven by the 'customer–contractor' principle, introduced in 1971 following the Rothschild Report (Gummett 1980). This is an administrative procedure by which it was hoped to make the work of the research councils more 'relevant' to societal needs. Instead of getting block grants to cover all their scientific projects, research establishments have to earn a proportion of their funds from 'contracts' for specific projects, commissioned by government departments acting as 'customers' for the research results. This procedure has been extended to cover most establishments in the public sector, many of which now have no core funding, and have to 'earn all their money' from a government department or other external sources.

Well-informed observers are not convinced that this new administrative procedure has lived up to its expectations (Kogan & Henkel 1983). The scientists affected by it are mostly very doubtful of its value. They recognize that it has had its effect on the pattern of research effort:

'It has come at a time when funds have been very restricted as well, so there is a smaller kitty, and one has to sharpen up projects. But, that apart, the projects that would have been worked on may have been rather different than the ones that are being worked on at the present time, because there has been so much emphasis on the direct practical application.'

but they remain sceptical of the ultimate value of the results:

'In some instances it produces much worse, less socially relevant, less useful, research than if you simply allow the scientist to go at it.'

This is as may be. For our present study, all that needs to be said is that the need to undertake *commissioned research* is putting R&D organizations under considerable stress. All the scientists with whom we talked were in agreement on the complications that it produces in managing research programmes:

'If you asked me what I thought of all this customer −contractor thing . . . I think it's terrible − a complete waste of time, and terribly bureaucratic. And I don't think it has led to any major alterations in the system . . . . [It] sounds fine when you hear it explained in theory, but when you hear how it operates − how these projects are tossed from one to the other according to how much money they've got in the kitty in any year − you see it being used as a financial device really. It's a nuisance, and it upsets people here because first they are commissioned, and then they are not commissioned, and they don't know what to make of it all. In fact, I can't remember half of which projects we've got commissioned or not.'

They are concerned at the sheer burden of getting funds. One scientist reported, typically, that when he wants to undertake a project that would formerly have been carried on the normal annual budget of the laboratory:

'Nowadays I would spend months trying to get money and approval, and get someone to give the financial backing to do it.'

Others noted the extra administrative work involved:

'In my own experience, it has not had much effect on the research which is done, but it has increased the administrative workload in justifying it, and getting the right money for it, and making sure that everybody knows what is going on, and having people come and check that you are spending all the money that you should be spending.'

These pressures are felt throughout the whole of an establishment. As a senior scientist put it:

'It's a pressure for change in the sense that the normal funding that you might expect to keep a good research programme going − some of that funding is being diverted, and has to be brought back by packaging up the research in quite a different way . . . . As a laboratory we have done quite well . . . and we have been left quite free. But the way things are going in terms of funding, and in terms of the restructuring of the academic work, the academic field, and so on, I don't think we can forget about those pressures any more. I think we have to attract something in, and it is a pressure. Maybe a few of us only feel it at the moment. It's lucky for the others, but . . . I think they will have to face up to the real world eventually.'

The effect is often to fragment the research, so that a particular group or unit might have as many as '20 different projects going all at once, some of them only lasting three months'. As we shall see in later chapters, commissioned research is often very different, in style, from the sort of research many scientists are accustomed to and calls for different motivations and different personnel policies. The recent increase in the proportion of this type of work in many R&D organizations is thus a very significant change from the point of view of scientific careers.

Above all, the present climate in the research world is of institutional insecurity. Some of this insecurity is a result of the application, from above, of general political doctrines, such as the efforts of the Thatcher administration to 'privatize' many government R&D establishments. Some of it is associated with the overall structural changes noted by Bailyn (1981) in the laboratories of firms in the US telecommunications industry:

A cycle that moved from research to development to production, with people at each stage working more or less independently, no longer fitted the new conditions. It was now necessary to make a direct transfer of technology, from research to production, and to forge much closer links between development and manufacture. But such changes affect the very essence of an organisation: they affect the meaning of time, the criteria for success, and the view of which people are most central to the organization's core mission. What was required was a cultural change, a term that was specifically used by the participants with whom I talked.

These institutional strains are often transformed into personal anxiety and anguish for the individuals involved:

'If your commissioned research or applied research funding ceases, you keep asking:'What happens next? There is not enough money coming in; where do we go from here?', and everybody feels very insecure about it. For instance, in my case . . . about four or five projects which were commissioned in the last eighteen months and finish on 31st December: all that money will have dried up. Where is my money going to come from: nobody said anything? I suppose there isn't any. I keep writing down, filling up

forms . . . Nothing happens, but you know one feels the sword of Damocles
is hanging there and it's going to go Boing! and that's it.'
In the next chapter we shall look at the scientists themselves, and the way in which
scientific and organisational changes affect their professional careers.

3
Researchers and their work

3.1 Research as a profession

Each year, thousands of young people are trained in some of the specialized subjects discussed in Chapter 1, and embark upon careers as research workers in the organizations described in Chapter 2. They acquire further skills, and eventually become fully-fledged professional researchers.

This book is about the effects of scientific and organizational *change* on the individual scientist in mid-career. As we saw in the previous chapter, the climate of the scientific world has altered a great deal in the past 20 years or so, and will surely continue to do so. But this general change is not a factor that individuals can easily take into account in their personal plans. Their long-term career expectations − cheerful or dismal as may be − inevitably presume a stable professional framework in which they themselves grow older and move on.

R&D activity goes on in an immense variety of organizations on a very wide range of subjects. Nevertheless, as we noted in §2.2, the work itself has many common features which characterize it as a distinct *profession*. But the scope of this profession is not easy to define, and it is not sharply differentiated from several others into which it merges. Thus, there is no public register of 'researchers'. Scientists usually have formal qualifications such as university degrees, but a minority of the people who get bachelors degrees in the sciences actually take up R&D work. Scientists do not have to be formally licensed, like doctors and lawyers, in order to practise their profession (§1.3). A higher degree, such as a PhD, carries considerable weight in the R&D world, but many researchers achieve high professional status without one.

As the term R&D indicates, it is usually impossible to draw a line where 'research' ends and 'technological development' begins. This, in turn, merges imperceptibly into the design, testing, and production of relatively standard products − that is into the normal activities of the *engineer*. In fact, as we shall see, many of the people working in R&D were trained as engineers, and continue to count themselves as such. Statistical data on persons employed in R&D often lump them together as 'QSEs' − *Qualified Scientists and Engineers*. Since other professionally qualified technical workers, such as veterinarians, dentists and physicians, are also engaged in the same activities: I shall follow the vulgar usage and refer to them all as 'scientists' or 'researchers', unless there is some reason to be more precise.

In some disciplines, such as clinical medicine, research work is often combined with professional practice. In some organizations, such as the Meteorological Office, people move back and forth between R&D work and the provision of a technical service to the public. The boundary is particularly vague in academia (§2.4), where research and teaching usually go on together, in the same buildings, often by the same people. Almost all academics are fully qualified and experienced in research, but they vary greatly in the amount they actually do. At one extreme, a certain number are employed as full-time researchers; at the other extreme, there are a great many teachers in universities and polytechnics who do so little real research that they should no longer be counted in the research profession (§6.5).

In later chapters we shall be particularly concerned with the very open boundary between the actual performance of R&D work and its administration and management. At what level of responsibility in an industrial firm or government establishment, does the 'bench scientist' become a 'manager' or 'administrator'? We leave this question open until we come to discuss the feelings that people have about making this transition (§6.1).

Throughout this study it will seem as if we are talking only about researchers in the natural sciences and their associated technologies, such as engineering, agriculture and medicine. This limitation is simply a matter of expediency, and does not signify any restriction in principle on the notion of 'scientific research'. It may be that full-time researchers in the behavioural and social sciences and humanities have special mid-career problems which deserve detailed investigation, but their numbers are so small within the research profession as a whole that it did not seem appropriate to give them special attention in this general study. Indeed, detailed studies would undoubtedly show that *every* scholarly discipline, from pure mathematics to management studies, has its own characteristic organizational structure and career problems (Whitley, 1984).

3.2   The demography of the research profession

How big is the research profession? How many 'qualified scientists and engineers' are there in the various sectors of the British R&D system? A recent official publication (R&D Review 1984) gives the following data for 'Manpower in R&D' in 1981:

Sector	Number
Academia	14,000
Research Council	7,000
Government	14,000
Private Industry	70,000

Research is such an ill-defined activity that these figures are not even reliable as 'round numbers. The figure of 14,000 for the academic sector, for example, is purely notional: it is made up by adding to 8000 full-time research staff an arbitrary fraction of the 19,300 full-time staff 'undertaking teaching and research in relevant subjects' (whatever that means!). Nevertheless, they do show that research workers now constitute a major professional group in society, comparable in size with such distinctive callings as medicine and coal mining.

From the point of view of the individual scientist, the total size of his or her professional group is less significant than its current rate of change. The research profession in Britain is obviously not expanding rapidly, as it was until about 1975 – but is it in decline? In spite of economic retrenchment in industry, education, and government services, the total numbers in the various sectors do not seem to have fallen appreciably in the past ten years. But these data are too coarse-grained to show what has been happening in particular organizations or specialized fields of R&D. As we saw in §2.5, there are pressures on many members of the research profession which have the same effect as a general decline in numbers. For example, the average mid-career academic scientist may not have been 'made redundant' (§9.1), but is expected to teach more undergraduates, and to make do with fewer graduate students, fewer research assistants, and poorer technical facilities, than was usual ten years ago.

The most significant demographic data are those relating to age and grade distributions. Here again, aggregate data often conceal serious unbalances. For example, the Holdgate Report (Holdgate 1980) showed that overall age distribution in the Scientific Civil Service was still fairly even in 1980, and that the median age had scarcely changed from what it was in 1975. But for a Principal Scientific Officer in his early 40s the significant fact was that the median age of staff in that grade had risen in just five years from about 46.5 years to nearly 48, due to the accumulation of a substantial 'bulge' of PSOs and SPSOs in their 50s, whose further promotion was thus blocked. Scientists in several establishments expressed anxiety about this sort of situation, which arose out of high levels of recruitment in the 1950s and 1960s.

In academia the expansion came slightly later, and has produced an even more extreme demographic unbalance. In 1979 – 80, in the biological sciences, physical sciences and engineering, the distribution of staff between ages 25 and 65 was approximately triangular, with a peak in the early 40s (Merrison 1982). In other words, even if there were no overall reductions in staffing, academics now in mid-career face the dispiriting prospect that there will always be fewer posts vacated by retirement than people fully competent to fill them.

A striking demographic feature of the research profession is the very small proportion of women, especially in the more senior grades. This is a complex issue, in which a variety of factors seem to combine to make it much more difficult for a woman than for a man to enter into a scientific career in the first place (Kelly 1981). These difficulties undoubtedly continue into later life. But so few women actually took part in our interviews that we have only scraps of anecdotal evidence concerning the particular problems that they have in mid-career. This is obviously a serious matter, calling for detailed research on its own account.

3.3   Organizational careers in research

A peculiar complication that arises in any discussion of scientific careers is the relationship between their 'reputational' and 'organizational' aspects. The traditional ethos of academic life (see, e.g. Ziman 1984) lays the whole emphasis on the *reputational* aspects of a scientist's career − subjects studied and mastered, contributions made to knowledge, recognition as an 'authority' in a particular field of research and symbolic rewards such as prizes or honorific office in a learned society. In other words, a career is represented as a path through various specialties (cf. §1.4) and through successive stages of esteem within the national or international community of scientists.

But scientists nowadays are seldom self-employed − like dentists or barristers, say − and they do not earn a living by reputation alone. As we saw in the previous chapter, they also follow *organizational* careers as employees of universities, government departments, industrial firms, etc. Such careers follow various patterns, but have many common features. It is uniformly accepted, for example, that one cannot enter into regular employment in research without having achieved a certain level of *higher education*, often followed by a period of formal *training* in specialized research skills. Many R&D organizations regard the first few years of employment as a period of *probation*, leading to a *tenured post*. The further careers of most scientists are punctuated by successive *promotions* to higher grades, often with heavy *managerial* responsibilities. Indeed, in many organizations research workers are simply fitted into the same career structure as other professional staff, such as administrators, accountants and middle management (§6.3).

The organizational and reputational aspects of a scientific career are, of course, mutually interdependent. For example, promotion to the professorial grade in an academic institution is earned by gaining a reputation as a scientific 'authority', and constitutes, in its turn, one of the tokens of 'recognition' of such authority. In other organizations, such as industrial research laboratories, the reputational aspect is still observable, but is probably much less significant. In fact, as we shall see in later chapters, the linkages between these two aspects of the scientific life are major factors in our whole theme (§7.5).

The organizational careers of scientists in contemporary Britain have a number of characteristic features in common. Any change in work roles is evaluated in relation to a stereotype which is not just a personal construct for each individual but is shared by all members of an institution or community (§9.3). The concept of a 'normal' career plays an important part in the way scientists think and feel about their work, although this is not always the way things turn out for them in practice.

The educational stereotype of a scientist is typically narrow in Britain. The choice of subjects leading up to examinations at 16-plus, and successful performance in those examinations, may effectively determine the field of study for the 18-plus examinations and thence into a specialized three-year course at a university or polytechnic. A 'good Honours degree' may then be seen as the signal, and the licence, for entry into a graduate course of training in research, culminating, after three or more years, in the award of a PhD.

This is the normal requirement for a career in academia, but it is not the only pattern. A considerable proportion of the qualified scientists and engineers in government and industry take up scientific employment after a first degree, and get their research training 'on the job'. Indeed, many R&D managers are scornful of the value of the conventional PhD training, and prefer to recruit graduates who are not, as they believe, too set in academic ways of thought (§7.2). This is a vexed question, which need not be taken up here, except to remark that there was no noticeable difference between the scientists in these two groups by the time we came to talk to them in mid-career.

In principle, it is not necessary to have even a first degree to do research. This is formally recognized in the Scientific Civil Service, which now has an integrated grade structure from 'Assistant Scientific Officer' up to the very top. In practice, however, it is not at all easy for a person recruited as an ASO, say, whose formal education has not reached degree level, to achieve the standing of a qualified researcher. Here again, we need not enter the debate on the merits of the 'Fulton' reform of 1972, which officially dismantled the formal barriers between the 'Scientific Officer', 'Experimental Officer', and 'Scientific Assistant' classes into which scientific civil servants used to be recruited. For our present purposes it is sufficient to note that it is unusual, but not abnormal, for a middle grade, mid-career scientist in a government or industrial R&D organization to have 'come up through the ranks', as one of them revealingly put it.

In the 1960s, when the research profession was expanding rapidly, it was quite normal for a scientist to go straight from the university into a permanent job in an R&D organization. Many fresh PhDs were appointed to university teaching posts where the probationary period was treated as a mere formality. This situation changed in the '70s. Many graduates are still being recruited by the major R&D organizations in the government and private sectors, but it is quite usual now for research scientists to pass through a succession of short-term appointments, in a variety of organizations, before getting a permanent job (§7.3).

Indeed, one of the most worrying features of the research profession nowadays is the tendency for this phase of a career to drag on into middle age. As a result of financial economies in all sectors of the R&D system (§2.5), a considerable number of well-qualified and competent scientists in their 30s – even in their 40s – are forced to depend for their livelihood on a succession of research contracts each lasting no more than one or two years. This is obviously a grossly inefficient way of getting research done, and there is very grave concern in the scientific world lest it should become institutionalized as a 'normal' career path. Nevertheless, it is at present a salient feature of the professional environment for a scientist in a supposedly 'permanent' job, faced with the prospect of losing it in mid-career (§9.1).

Once established in a large R&D organization, a scientist would expect to stay there for the rest of his or her career, moving up from grade to grade. Every organization has its own hierarchy of ranks, grades, titles, and posts, but these mostly follow the same general pattern. For example (Fig. 2), the Scientific Civil Service is stratified into eight grades (excluding three more higher grades in the 'Open Structure'), within which it would be normal to rise by about three grades in the course of a career. But most of this promotion comes early on. Thus, a promising young scientist with a higher degree, recruited as an HSO at the age of 25, might reasonably expect to be a PSO at about 40, but could not be quite sure of further promotion to SPSO before 50 – if at all. Of course, a few 'high flyers' rise much more rapidly, to become directors of establishments or senior officials in Whitehall. In general, however, the age–grade distribution in most R&D organizations reflects – and constrains – a much more modest career path for the great majority of their staff.

The more senior grades in most R&D organizations – e.g. SPSO and above in the government and research council sectors – are primarily managerial or administrative (§6.2). But there may also be a 'dual ladder' of senior grades without such responsibilities for a few outstanding researchers – for example,

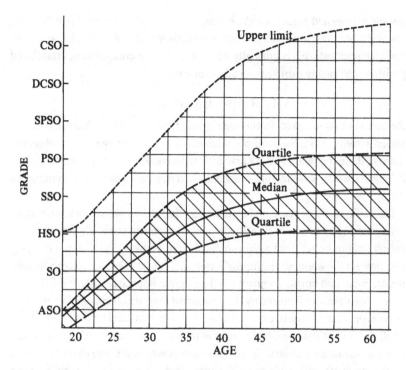

Fig.2 Age/Grade distribution in Scientific Civil Service (data from Holdgate 1980).

the *Individual Merit Promotion* (IMP) scheme for the permanent scientific staff of government departments and the research councils. This ladder is thus designed to parallel the hierarchy of posts in academia, where the title and emoluments of a professor are awarded primarily for scholarly performance, and only incidentally carry managerial responsibilities (§5.3, §7.6).

It is sometimes remarked of a very able scientist that he or she 'never really retired'. Nevertheless, it is a fact of life that an organizational career must be formally terminated at a certain age – usually between 60 and 65. This coming event tends to cast its shadow over most people in their late 50s, strongly affecting their attitudes towards organizational or scientific change. Some of the factors entering into these attitudes are connected with specialization, but they are compounded with other factors which lie outside our present study. The 'pre-pension' period is evidently a distinct career phase which needs to be studied in its own right (Sofer 1970).

The 'mid-career' phase is thus quite a long segment of adult life, from the end of the educational and probationary phases to the pre-pension phase – from 35 to 55, say, in the normal pattern. For the few who have already been singled out for high preferment these are likely to be exciting and strenuous years with great responsibilities and great rewards. For another atypical group, they may be years soured by personal misfortunes lying outside their work. But for the majority of research scientists this is normally a period of proven competence, established position, and steady professional achievement.

3.4   Research as a vocation

Research is an ill-defined, institutionally fragmented, and technically extremely varied activity, but it has certain characteristics which unify it as a profession. Some of these characteristics are practical, and have already been noted in Chapter 2. There are also less tangible conceptual and psychic features which make it a distinct *vocation*.

Scientific workers share widely the belief that all scientific work is uniquely characterized by a 'scientific attitude' and a 'scientific method'. Whatever the validity of this belief from a philosophical point of view, it provides a unifying *mystique* for the whole scientific profession, and can be linked sociologically with the practices and norms common to all academic science (Ziman 1984).

Scientists are also inspired by the notion that they are all engaged in the same general process of 'discovery' − a unity that is emphasized by the involvement of all fields of science and technology in the same institutions of higher education and advanced scholarship. All researchers nowadays also share the notion that they contribute directly to society through their 'problem-solving' role − a role that is underlined by the intermingling of 'scientific' and 'technological' goals and methods in the sponsorship and performance of most forms of research.

In the present work, we are particularly concerned with questions of personal *motivation*. An outstanding characteristic of research is the extreme satisfaction that it apparently gives to those who do it. Scientists may become discontented and divided about their terms of employment, their promotion prospects, the way their institutions are managed, and other aspects of their organisational careers, but they are united in their personal enjoyment of the work of research. In the interviews, they continually confirm the observation of Eiduson (1973):

> 'There seems to be almost no activity for the scientist that offers so much gratification as science: it is no wonder that other activities and activities with others are pale by contrast.'

Why is science so gratifying, even for those who are not strikingly successful at it? It is not as if there was much social esteem to be got from scientific research, on a day-to-day basis. Scientists do not have much opportunity to help their fellow human beings in trouble, like doctors and nurses, or boss them around, like army officers or workshop foremen, or win contests with them, like barristers or athletes. Few R&D projects call for such 'urgent' (§2.3) responses as in journalism or politics:

'[In my Final undergraduate year] I realized I had a much lower boredom
threshold than I thought I had actually got, and could not see myself sitting,
sweating over one problem for three years, and realized, having only spent
one year in industry prior to going to [university], that was the kind of area
I wanted to work in – an area of flux and challenge, where two days are never
really the same.'

Superficially, research is a quiet, and often solitary *craft* (Ravetz 1971). A series
of elaborate technical tasks have to be carried out, skilfully and carefully,
according to a predetermined plan. These tasks are often highly standardized and
quite familiar, even though they may require years of training. 'Ethnographic'
studies have emphasized the everyday ordinariness of 'laboratory life' (Latour
& Woolgar 1979). In every field of research, however 'exciting' the discoveries
that are being made, there is always an enormous amount of routine technical work
to be done.

Some scientists undoubtedly get their satisfaction from the sheer craftsmanship
needed to do their job well:

' . . . if you feel you've achieved something then you get satisfaction, even
if it's flying around in an aeroplane rather than appearing in a learned journal.
You can get the same sort of satisfaction'.

But the scientific workers with whom we are mainly concerned in this study are
not just technicians. In the longer term, their work is certainly not 'routine'
(Hargens 1975). A research project is not a programme of specific tasks with fixed
deadlines, and its problems, concepts and techniques are always changing (§1.7).
The challenge of continually solving new technical problems is evidently part of
the attraction. Scientists speak of being 'fascinated with [their] research topic' and
'totally immersed in looking at it'. One scientist spoke of the pleasure of making
a discovery:

' . . . the thing that gives me the kicks are the things that I would call intellectual orgasms, which I have once every blue moon, where I find out something
which I regard as new and interesting, and suddenly it comes'.

How far is this satisfaction in solving a problem linked with *autonomy* in the choice
of problems to be tackled? Researchers continually emphasize the value they
attach to freedom to 'do their own work':

'I've been very lucky, in the sense that I've never been asked to do anything.'

– a freedom that was even permitted at one time in an industrial R&D laboratory:

' . . . I went straight into basic research, and I was never told what to do.
I don't think I ever received an instruction what to do when I came, but I had
some ideas on . . . . There was little relationship really to the company's
products'.

Personal autonomy in the performance of one's work is the hallmark of any established profession. But there is an important difference in principle between *strategic* autonomy in setting the problems for one's research and *technical* autonomy in attacking and solving them (Bailyn 1981, 1984: Whitley 1984). Strategic autonomy is often held to be an inseparable element of the *commitment* (Becker 1960) that scientists are supposed to give to their work:

> 'When you are just doing a job that people have found you, as opposed to a job that you've got yourself heart and soul into, it's like being on an assembly line of a car industry. When your heart and soul is into a job, five o'clock comes and most people do go home, but they don't stop thinking about it.'

One researcher expressed this sense of commitment even more strongly:

> 'And thus was born a sort of uncontrollable passion, really, for the subject, which has carried on all the time.'

Many, but not all, scientists would probably echo this sentiment: we shall come back again later to the important question whether autonomy in problem choice and commitment to a particular subject specialty are indispensable vocational attributes of the research profession (§9.3).

Nevertheless, a common characteristic of the research profession is its idealisation of the researcher as a 'self-winding' person:

> 'Yes, self-motivation I think is the most important factor, whatever way you're going, and self-satisfaction feeling that you've done a good job; and then, after you feel you've done a good job, you feel you would like to be recognized for it; but that comes, I think, the next stage on.'

It may be, as Eiduson (1973) has deduced from psychological tests, that this is 'an overdetermined reaction, because the scientist feels that he knows and can trust his own personal resources, but he has doubts about relying on others and trusting them'. All the same, this is a sentiment that was apparently shared by all the cientists we interviewed.

Although scientists share a general feeling that science is of benefit to humanity, they are not strongly motivated in practice by consideration of the longer-term applications of their work (Jagtenberg 1983). In academia, as might be expected, commitment towards the instrumental function of science is only lukewarm:

> ' . . . even though there isn't any constraint to do applied work, I think that most of us feel . . . that if we had applied work that we could do we ought to be interested in doing it . . . because we are an applied department.'

But even in industrial research, which is organized entirely for use, it is treated as boundary condition rather than a primary motivating factor:

> ' . . . a chap who works for me has a fairly wide area in which he works, but even so the end product of his research has to be something that can be

used operationally . . . he's aware of the environment into which his work fits, so he works towards a product which he knows would be used in a certain way'.

The impression one gets is that social use is considered a public, institutional value, rather than a source of personal gratification:

'Universities lay great stress on doing pure research. My phrase is ''Conquistadors of the Useless'', which I have been dying to use . . . this man is just climbing for the sake of it: he loves climbing, he set himself a task, and he gets his kick out of doing that . . . in universities great stress is laid upon conquering something that is totally useless and has no applicability to the general human good.'

If scientists are, in fact, motivated altruistically they are seldom explicit about it. It was rare to hear remarks such as:

'But to some of us, you know, the actual value of our work in terms of implementing, serving the community, the farming community, which, after all, we are here for, is a prime consideration and motivation. If this is not satisfied . . . you find yourself getting rather jaded in attitude.'

3.5 Personal effects of institutional change

The institutional stresses described in the previous chapter (§2.5) are producing radical changes in the research profession. Scientists nowadays are often uneasy about the public standing of their work:

'In the past, science was − I think certainly in the 1950s and 1960s − a very respected, highly respected profession, with a lot of prestige. A lot of potentially good people were drawn in. But now it's a dirty word. It's dropping down in the standing of most people as a career, as a job.'

This supposed decline in public esteem goes hand-in-hand with anxiety over the loss of the many employment openings, the promotion prospects and the job security that used to be so characteristic of science as a career. We have already noted (§3.3) that a great many fully qualified and mature researchers are now being employed on a succession of very short contracts without the prospect of getting tenure within a few years. Because of their highly specialized employment, the R&D staff of a company may be the hardest to reduce by wastage, voluntary redundancy and early retirement (Hutt 1981). In all sectors of the R&D system there are now threats of redundancy for competent mid-career scientists whose jobs had always seemed perfectly secure − and the knowledge that there are often hundreds of well-qualified applicants for the few comparable jobs that get advertised.

For the great majority of mid-career scientists the risk of being thrown on the job market may still not be very grave; but their prospects of promotion have undoubtedly become very much worse in the last few years. As we saw in §3.2, the abrupt transition from expansion to level funding in the mid-1970s left the whole research profession with a great excess of older people who will take many years to pass through to retirement. The anxieties of those whose career ambitions are blocked in this way communicate downwards to the younger people, of whom there are not proportionately so many as in the past, and who are said to be 'under increasing pressure to earn their keep', and 'not generating the ideas' that they used to.

This atmosphere of disquiet is having complex effects on personal motivation and achievement amongst scientists. The conventional wisdom of everyday life is that security breeds laziness and orthodoxy, and that a certain amount of insecurity is necessary as a spur to individual exertion and a challenge to the creative imagination. Many people now consider that the research profession should not be excluded from the traditional managerial doctrine that one must 'treat 'em mean to keep 'em keen!'.

On the other hand, security and stability of employment are highly valued in science, not just for themselves but as the framework within which people are willing to take substantial long-term risks in their research (see §9.1). It is generally held that scientists who are free to choose their own research problems, but who are not quite confident that they can keep their job, or get another one without difficulty, are tempted to choose 'safe' subjects:

> 'If you reduce the feelings of security in scientists in general, in other words, if you start them panicking, then the research changes very markedly. You can see this in the difference between people with tenure and without tenure. To my mind it changes very much for the worse, and people start to do research which they know will get results, and get papers, and it's not the sort of research which in general is conducive to progress in science.'

This opinion, which is supported by objective evidence (Lemaine *et al.* 1972; Reuter *et al.* 1978), indicates that insecurity and enhanced competition do not produce their benefits automatically. They can only be gained if the opportunities for imaginative change in one's personal endeavours do not seem too risky. This is a key point in the present study.

As we saw in §2.5, the policy of all R&D organizations has shifted significantly towards more 'urgent' and 'relevant' projects. Scientists are not against this in principle, although concern is expressed at 'using up the seed corn', and of no longer 'being in the forefront' of basic science. What worries them much more is the way that research programmes are now being funded:

'There are two different sides of research. One is the side of research which is the intellectual excitement. You are following a problem because it is interesting, and if somebody asks you to prepare a five-year forward look you would say, "This is impossible, because I don't know where this work is going to take me during the next 12 months, and if I gave you a five-year forward look it would probably be wrong, I hope, because exciting things turn up" . . . . But there is the other part of science that requires you to justify your work in terms of funding, which means that you slice the work into easily digestible lumps which you think people will buy . . . You give a forward look for three years, and say "I will do this, and if you pay me for it I will do it, and give you the reports on 30 March 198 – , which will contain all the information listed here . . . . Those pressures are on us now, so the nature of science is being altered . . . ".

In the research council and governmental sectors, this change in the nature of scientific work is often identified with the advent of the 'customer – contractor' system for programming research (§2.5). The 'customers' who 'commission' the research are said to be ignorant of their real needs, inflexible in their demands, and not interested in 'feedback' as the research proceeds. In addition, many scientists complain that such research is excessively fragmented:

' . . . you have half, quarter, a third of a person on a project . . . . I spend a quarter of my time on about four different activities, and find myself doing none of them . . . .

We just spend one day on one project and then the next day on the next . . . '.

and that they have to spend far too much time writing progress reports which are not worth publishing.

Such complaints may well be justified, but they should not be seen as directed solely at the administrative procedures of the 'Rothschild' system. Similar complaints may be heard in academic and industrial R&D organizations, for they are symptomatic of the changed circumstances of scientific work in general. They refer, in particular, to the way in which the individual researcher is being forced to cede control over his or her personal research programme to an external 'customer', to an anonymous 'peer review' panel of a grant-awarding agency, or to a cost-conscious company management. Instead of working at their own pace, at their own chosen problems, in their own little groups, scientists are having to fit in more with the work of other people, in other organizations, with other objectives.

At the worst, this may mean having to take up some entirely new and unfamiliar topic, in order to move to another firm, or get posted to another division of the establishment, or simply to win a research grant for one's group. It may mean having to put aside research that looked promising and fruitful in order to answer a much less 'interesting' question posed by a 'customer'. It may mean going out into industry, or peering into the recesses of Whitehall, to seek commissions and negotiate contracts to keep the establishment going.

These are the changes in the practice of their profession that many well-established scientists find difficult to adapt to in mid-career. Having built up a reputation as a subject specialist (Chapter 1), a scientist can be as fearful as any miner or steel worker of being completely 'deskilled' by having to move into a different specialty (§5.3, §5.5). This concern applies even to doing commissioned research in one's own field:

' . . . I don't think you are using all your skills, because some of your initiative is taken away. Some of your own individual insight is taken away . . . you have to restrict your own inquisitiveness . . . . So you are only using part of your training . . . part of your abilities'.

The difficulty in 'doing what people pay for in a way which stimulates you' is not mere prejudice, for one may know that the results produced will not help to 'build up a reputation which stands me in good stead when I next meet my promotion board' or feel that the project one is engaged in 'does not equip me well to compete for senior scientists' jobs in the Civil Service'. (§7.5). They say (not without reason, in some cases):

' . . . a lot of these commissioned projects, you don't actually want your top university scientists, you want another type of person altogether'.

3.6   Adaptation to change

In every profession, however thriving, a certain number of individuals come to feel that they have made the wrong choice of career. But in the scientific profession, which was previously notable for the satisfaction it gave to its members (§3.4), the disaffection is now more general. Many competent and mature professional researchers would agree with the scientist at a quasi-academic research council establishment, who summed up the situation as follows:

'There has been a very real change . . . in the kind of job we are doing. When I first came to science, I thought, from the people I worked with at university and from the kind of place this was, that I was going to do pure research, and I would, as long as I continued to do this, presumably to the satisfaction of somebody – that would be my career. If I had wanted to do work which was directed, I think I would have considered going into industry, with a different kind of structure, and a different kind of career organization altogether, with a view, perhaps, subsequently, of moving through industry in a different way. I think this is perhaps the change, the change of the environment in which we find ourselves. I am not saying that personally this has happened to me yet, but the whole environment in which scientists find themselves in academic institutions is changing.

The reasons, I think, may be very good. It may be reasonable for people who pay the bills to have some say, but this is very different from the kind of world we entered. When I entered it in 1967, it was a different kind of world, and now, you may say, "Well, that world should perhaps only exist in universities and not in research institutions". But then, we could say, "Well, now, unfortunately, going to a university is virtually impossible, because universities are shedding staff, not recruiting them, and they too are under these pressures". And that is the change: it may be a real change to expect, but it is a change we have had to experience.'

This decline in morale must be seen in perspective. A number of institutions have carried their staffs along with them in the transition to new modes of programming and funding research:

' . . . we have been fairly successful at establishing an atmosphere now, a general atmosphere . . . where considerable flexibility in following new opportunities is taken for granted and applauded, and people therefore . . . gain the respect of their own groups — of this community as it exists now — by success in application, rather than in pure scientific achievement and publication as such'.

Individual scientists have learned that:

'there is a lot to be gained from commission research . . . You are doing things which need to be done, and getting paid for it. More important, as far as I am concerned, if we arrange these projects properly so that they are not just number-producing — so they are fully fledged research projects — it gives you access sometimes to material problems which you would not be able to get to otherwise.'

Others evidently get some satisfaction from their new functions:

' . . . actually going out to industry or to requirements, boards, and understanding what the customers require of the work, the degree that they are actually prepared to pay for it, and then coming back here and finding people, outside the main management-line area, who have the capability — can be manipulated into actually doing the job'.

In other words, many scientists are effectively adapting the new organizational structures to their own ambitions, or making new careers for themselves within them.

In any case, the majority of QSEs are accustomed to working in large R&D organizations where the limited autonomy and other effects of 'collectivization'

(§3.1) have always been regarded as normal. Some voices outside the research world (§9.1) even advocate a ruthless policy in which the disaffected or inflexible minority in the university, research council and government sectors would be replaced by new cohorts of scientific workers, trained in and accustomed to contemporary conditions of R&D activity, who would probably carry out their duties cheerfully, confidently and quite reasonably competently.

But such a policy would impose terrible costs on the scientific community, and on the nation. From a strictly utilitarian point of view, the thousands of professional research scientists now in mid-career constitute a national resource that should not be simply scrapped just because their skills and attitudes are thought to be obsolete. More importantly, these are a group of able and sensitive people, each of whom deserves to be helped through a personal situation that is not of their making. 'Redeployment', rather than 'redundancy', is surely the wisest and most humane policy.

The real question, then, is whether the radical transformation of many aspects of the research profession is imposing intolerable demands on many scientists who are already in mid-career. Can they be helped to adapt themselves to rapid changes of research programmes, research techniques, and organizational roles, in response to new 'customer' needs and new institutional structures, as well as the accelerating advance of knowledge itself? These are large questions, which can only be answered by reference to all aspects of the management of R&D organizations.

In the present book, we concentrate on just one of these aspects. As we saw in Chapter 1, the primary fact of all scientific work is specialization. The division of the labour of research into narrow specialties is unavoidable, but it does present great obstacles to change, both for institutions and for individuals. What are the real difficulties encountered when research scientists are impelled, for one reason or another, to move out of their individual subject specialties (Chapter 1), and how can these difficulties be met and overcome?

4
Versatility

4.1 Adapting to abrupt, involuntary change

We have now arrived at the heart of our investigation. The majority of scientists, when left to themselves, normally choose to work for years at a time in quite narrow fields of research, with only gradual changes of topics or techniques (Chapter 1). But as we saw in Chapter 2, R&D organizations are now expected to respond rapidly to external forces, and to modify their research programmes accordingly. Scientific researchers are thus coming under pressure to change to new fields at a much higher rate than they would have done of their own free will (Chapter 3). Scientists are accustomed to making *slow, voluntary* changes in the subject matter and methods of their research: how do they get on when they are expected to make *abrupt, involuntary* changes in their individual research projects?

Let us first deal with an obvious point of principle, which affects all studies of human behaviour. It is a matter of common experience that people seem very variable in their innate capacity to 'adapt' to radically-changed personal circumstances. The ease with which a particular person is able to move into an unaccustomed situation and perform competently therein is often ascribed to a specific personality trait of 'adaptability', connected with other traits of intelligence, emotional stability, etc. (§4.8). But common experience also teaches us — as in wartime, for example — that this trait is often 'latent' — that people can 'rise to the occasion' and 'face up to the challenge' of a radical change of circumstances in ways that neither they themselves, nor other people, would previously have suspected. In other words, the whole issue we are studying is bedevilled by indefinable and indeterminate factors that are highly subjective, and vary greatly from individual to individual.

But within the complex of thoughts, feelings and actions associated with such life events one can distinguish certain motivational factors that can be explicated relatively objectively. A number of quite general and perfectly rational considerations arise, for example, when a scientist in mid-career tries to estimate the effect of such a change on his or her further career prospects: these will be taken up in Chapter 5. Such considerations, in turn, relate to the realistic possibilities of moving out of research along other career paths, such as administration or technical services, which will be discussed in Chapter 6. From these discussions it will be apparent that the 'adaptability' of researchers really depends very much on specific organisational policies and institutional practices: these will be dealt with in Chapters 7 and 8.

But motives can only be analysed rationally within a realistic framework of capabilities. People cannot reasonably be expected to take up difficult jobs for which they have no competence. Scientists, in particular, establish themselves professionally by their expertise in very narrow subject specialties (§1.3): to what extent would they be *technically* competent to do research in somewhat different specialties? How widely applicable are the skills and knowledge that they already have? How did they actually acquire this expertise? How much of an effort would it be for them to get to know the basic principles, the standard techniques, and the outstanding research problems of a relatively unfamiliar subject? In this chapter, we shall be concerned with questions such as these — that is, to use another ill-defined, everyday term, how *versatile* are research scientists?

4.2   Regions of versatility

The first question is: what are the *limits* to the versatility of a professional scientist? The most striking cases of personal adaptation to enforced change occurred during the Second World War. Not only did bank clerks and solicitors end up as colonels and sergeant majors: more surprisingly, a number of academic zoologists and botanists quickly became brilliant radar engineers or operations analysts [?]. Efforts were made to recruit scientists into wartime work that was relevant to their previous experience, so such cases were not very common. Was it only a few outstanding individuals who were successful in making such radical changes of field, or do most scientists have this capacity? Unfortunately, the historical evidence from the War period is too fragmentary and anecdotal to answer this question convincingly.

The scientists who took part in the group interviews were of a younger generation, and had never been under such pressure to work in unfamiliar fields. Nevertheless, a few of them had, in fact, made large changes of specialty in the course of their careers. As we saw in §1.6, such 'migrations' across the cognitive map are not very common, but they do indicate the extent of the region within which a competent research scientist can actually move without having to start again almost from scratch.

These cases support the natural assumption that there should be no insuperable barriers to the migration of a competent scientist from specialty to specialty within a conventional academic 'discipline' (§1.1). For example, when a particular establishment needed to replace a radio chemist, it was rightly considered that 'a good chemist' could adapt quite well to this specialized job. Again, physicists who had worked mainly on the release of fission products in a nuclear reactor found themselves quite competent to take up the totally different problem (from a physicist's point of view) of the effect of an explosion on the mechanical structure of such a reactor. It was not surprising to hear that a zoological neurophysiologist could have switched, if need be, from a dogfish to a dog as an experimental animal, or that a botanical biochemist had moved from a marine fungus to a cereal. As an agricultural researcher put it:

' . . . we are into agriculture in general, and not into, shall we say, virus transmission *per se* . . . . The fact that – has moved from soft fruit to cereals really does not matter too much because the general principles are the same. People learn a new set of parameters, and there is a period of familiarization. I think those sort of transitions can be made relatively easily . . . . It's still pathology, and the basic principles of pathology hold.'

There were suggestions in the interviews that such a 'region of cognitive versatility' may actually be larger than a conventional academic discipline. For example, many biological mechanisms are almost universal, so that the whole of biology is often considered accessible to a knowledgeable biologist. Thus, the boundary between zoology and botany was easily crossed by the researcher whose study of cell division in a human embryo had turned out to be an adequate basis for starting research on the growth of the roots of plants. Similarly, moving along another axis of the cognitive map of biology, a parasitologist could become a toxicologist in a relatively short period.

A correspondingly large region of versatility in the physical sciences was indicated by a scientist whose present work might best be defined as chemical engineering:

'I really feel that there is so little difference between physical scientists and engineers . . . . I can tell the difference between biologists and maybe metallurgists, but between electrical engineers, mechanical engineers, civil engineers, physicists, chemical engineers, if you like aeronautical – they are such fine divisions that I think they are almost interdenominational. I will put it that way. You know – they can put their collars on backwards sometimes, if they really want to.'

Thus, scientists themselves do not accept the conventional division of the scientific map into mutually exclusive disciplines. In many academic institutions, the sciences and their associated technologies are grouped into very broad transdisciplinary 'schools', under headings such as 'The Physical Sciences', 'The Biological Sciences', 'The Earth Sciences', 'Materials Science', 'Biomedical

Science' 'Engineering', etc . . . . These groupings cannot, of course, be defined precisely, and they overlap one another in complicated ways, giving rise, for example, to interdisciplinary subjects (§1.2) such as biophysics or pharmacology. But within each group one can usually find a common body of elementary facts, concepts and methods – that is, a common scientific 'language'. The thousands of scientists working in the hundreds of research fields within such a region may speak very different 'dialects' of this language but they are able to communicate with one another about their research, and each can understand, at least in broad principle, the problems being investigated in other fields.

   This notion of the sciences falling into characteristic 'language areas' is obviously very schematic and imprecise, but it does suggest distinct limits to the cognitive versatility of a mid-career scientist. In particular, any field of research where the standard 'language' is essentially mathematical is almost impenetrable without a formal mathematical education. For example, the fact that 'a lot of geologists . . . are just not used to thinking in terms of numbers' clearly restricts their versatility in this direction. On the other hand, physicists and engineers can at least understand the mathematical formalism in a field such as ecology, even though they may have a great deal to learn about the reality to which this formalism is applied. Interestingly enough, the mathematics barrier does not seem to include direct computer modelling: as we shall see (§4.6), scientists with negligible formal mathematical education can often pick up the basic skills of computing and apply them very effectively to their research problems and techniques.

   The immediate difficulty in getting to understand a different scientific language is in mastering a new vocabulary or a new formalism. Thus, for an applied mathematician having to do research in geology, this meant:

   ' . . . mainly learning the jargon, as it were, as far as we are concerned. We have to understand geological reports. You've got to learn how to read geological reports, which is no mean task unless you are a Latin scholar. But the other thing is three-dimensional visualization'.

But the deeper difficult is in learning to adopt a different point of view – even a different philosophy of research. This may have been the problem faced, for example, by an ecologist forced to bring economic factors into his research:

   ' . . . I would try to understand the principle of discounted revenue, but how can you incorporate discounted revenue into assessing farm values, compared with forestry values? That is why I am struggling, and that is where you've got to rely upon people who understand these things far better than I could understand them'.

Similarly, an experimental physicist remarked that:

   ' . . . people trained in exact science in a complicated, complex system of biology find themselves in a very different world, and would probably find it very difficult to adapt [to our work], as we would find it very difficult to adapt to doing biology, where the systems are notwell-defined'.

This is confirmed by a detailed study of a French biological research unit (Lemaine *et al.* 1972), where 'many of the researchers, being physicists and chemists, had to go through an intellectual reconversion, which turned out to be particularly difficult for the physicists, who had to learn to ''lose'' some of their rigour and their fundamental concepts.'

From this discussion it might be thought that in a science such as physics there would be a barrier between the experimental and theoretical branches, whose research techniques are so different that they are sometimes regarded as distinct professions. This does not seem to be the case. It may not be easy for an experimental physicist to make a contribution to the more esoteric aspects of mathematical physics, but he or she is usually quite competent to understand and analyse theoretically the results obtained in a particular problem area. Conversely, a researcher who began essentially as an applied mathematician may become involved in 'planning and organizing experimental work', even if they do not actually have to set up the apparatus and twiddle the knobs until a result is obtained.

Similarly, there does not seem to be an insurmountable barrier between the 'pure' and 'applied' aspects of a science. Indeed, a general movement towards more 'urgent' and 'relevant' projects (§2.2) is typical of many scientific careers. For example, there did not seem to be any intellectual difficulty when a scientist:

'. . . decided . . . that I should try and use the knowledge that we gained to try and solve applied problems, particularly overseas, where really one is using very much simpler theoretical models — much more empirical models . . . '.

— even though, in other respects this was clearly seen as a major change of career path. In some cases it was even recognized that personal movements back and forth between a 'pure' science and its related technologies was a valuable experience for the researcher, and helped to diffuse consciousness of new discoveries or urgent practical needs within a particular 'language area'.

4.3   The effect of education

Subject specialization begins very early for most British scientists (§3.3). By the age of 15, they may already have had to drop one of the standard school science subjects — physics, chemistry or biology — in order to keep going with mathematics, and with a minimum of humanistic, linguistic, or social studies. This specialization is further narrowed between the 16-plus and 18-plus examinations, during which time they will study only science subjects and most biology 'specialists' will cease to learn any mathematics. At a university or polytechnic, the typical 'Special Honours Degree' is (as its name implies) designed to carry the student as far as possible towards the research frontier in only one discipline. Is the narrowness of their formal education one of the reasons why scientists and engineers do not adapt easily to change?

In the discussions, each scientist naturally referred back to their first degree course in a particular discipline, and usually commented on its relevance to their subsequent careers. But they seldom saw their school and undergraduate education as highly specialized. On the contrary, whatever the discipline, it was looked back on as a source of rather general and basic knowledge. A variety of remarks make this point quite clear:

A. ' . . . when you are doing a [modern] degree in chemistry you come across such a wide . . . range of things. You don't just sit the whole time over test tubes. It's such a very wide subject. You use the computer — all the varied appliances.'

B. 'Not so much as geology. Geology is perhaps the classic example of the degree where you need a versatile education before you go into it, because you're asked A levels in Maths, Physics and Chemistry — the basic scientific subjects. And it's the way it's treated all the way through: all different sides all thrown together.'

'I'd been told — deceived — that a general degree like physics would be very appropriate for a whole range of possible developments later, and while at the time I was interested in particular areas, as, for example, if I'd done electrical engineering, so I made the decision to stay with the general subject . . . .'

Characteristically, they see *other* disciplines as narrower than the one they were themselves trained in, e.g.:

' . . . physics is a very good training because in a way, in physics you learn a little bit about so many different things. You know, it's optics, and it's electricity, and it's heat, and it's so many other things, whereas some of the engineering people that I do know — they are rather narrow . . . they have never looked at optics, or something like that.'

Another member of the same discussion group gave a different view —

'I am an engineer, but then you see I was rather fortunate . . . . The course I did was a sort of very broad, very broadly based thing, and we did quite a lot of physics . . . . Though it was mechanical engineering basically, we did physics, much more than one would normally do in a mechanical engineering course, and a considerable amount of [other] engineering as well.'

The point is, of course, that we are dealing here with mature scientists, whose undergraduate training lies some 20 years behind them. During that period they have mostly been working in research specialties whose extent may have been no more than a few per cent of the extent of an undergraduate course (§1.3). They have had to learn so much more about their particular specialty, either in formal postgraduate courses or by individual effort, that the few lectures they had on it as students now seem very elementary and trivial. A physicist who 'had no problems at all in becoming a chemical engineer' made this point very confidently:

'I think you only use about 3 – 5per cent of your undergraduate training at postgraduate level, anyway, even if you stay in nominally the same field of of physics, and I think if you've got a scientific and technical training you can pick up the 3 – 5per cent in another scientific or technical field . . . very quickly indeed, in a matter of months.'

In any case, what they learnt as students was of limited scientific value until it had been put to use:

'. . . when you first leave, say your first degree, most people then probably know more ''geology'' than they ever [do later] . . . . You know more facts as an undergraduate. You don't understand them . . . until you have some real experience.'

Their formal training thus fades from memory as it is overlaid by all they learn about science in the course of doing research:

'I think if somebody asked me some problem about the geology of – I would not start to go back and think of things I knew as an undergraduate. I would probably start to think in terms of the physics of the problem, or something like that . . . or the chemistry . . . . I would start from scratch. If I needed some facts I would know where to get them. But my approach would not be the same as it would be, or would have been, just by using undergraduate [knowledge].'

As academics themselves insist, the function of even the most specialized Honours degree is not to produce ready-made research specialists, but to introduce students to the basic elements of a scientific discipline. The aim is to teach them to 'think physically', or 'chemically', or 'geologically' or 'biologically' – that is, to train them in the 'language' of a particular branch of science (§4 2) rather than in its detailed results. As a scientist who had actually moved away from research into technical marketting remarked:

. . . the fact that you've had the background of physics training and you have done work in it gives you an enormous degree of subconscious core. You don't have to think about a lot of judgements because you just know . . . '.

It could be argued, indeed, that modern science degree courses *are* too specialized, in that each of them may cover only a small part of its 'language area' (§4.2). But academic disciplines are not sharply bounded. and research scientists are not professionally constrained to remain within the field of their first degree (c.f.§3.1). For example:

> '[I was] studying chemistry at — for three years, and during the final year I got more interested in the lectures on biophysical chemistry. Staying within that department, I went and did a PhD in biophysical chemistry, and so I was working with viruses and bacteria . . . which really started me off on the track towards biology.'

Again, a scientist whose training was in geology was now 'falling into materials science', another who had taken a degree in meteorology was able to take a postgraduate course in optics and move into applied physics, whilst another remarked that:

> 'the language of, shall we say, a degree in applied mathematics was sufficiently close to the language of a degree in engineering that, once you had been forced to do experimental work, you soon pick that up'.

By the time that a scientist has reached mid-career, he or she may have drifted such a distance across the landscape of science (§1.7), or that landscape itself may have changed to such an extent (§1.2), that the label of a first degree may have become quite meaningless. This was particularly evident in fairly new interdisciplinary subjects. A scientist remarked that when he joined a particular establishment:

> ' . . . what I thought were radio chemists turned out to be part botanists and part plant physiologists and plant biochemists'.

At another establishment, a 'discipline group' on materials included metallurgists, chemists and physicists. In a group of ecologists, one of them pointed out that:

> ' . . . very few of us were trained as ecologists. Most of us have been trained in quite different branches of science . . . We are now calling ourselves ecologists, where previously we called ourselves botanists, foresters, physiologists . . . '.

As it was put most forcefully in another context:

> ' . . . the idea of branding people physicists, or chemical engineers, or mechanical engineers, or whatever, because that is what they did as an undergraduate, or even at graduate level, is a nonsense as far as I am concerned!'.

All this is not to say that the degree of specialization in the education of scientists is not a matter for serious concern. There are good reasons for believing that this specialization is excessive in the last three or four years of secondary school, and perhaps in the earlier stages of many degree courses (Ziman 1980). But the damage it does is in the limited education of young scientists in matters outside of science, and in the false impression that they get of the scientific enterprise as a whole, and of its place in society. The technical versatility and adaptability of scientists in mid-career depends much more on their personal experiences in their professional work than on the subjects they hazily remember from school or university.

4.4 Acquiring sundry skills

Scientific research is not *routine* work (§3.4). Every project is to some degree new, and demands new information, new ideas, or new techniques in order to get new results. The researcher is always having to learn something new. The technical and cognitive skills that a scientist may thus acquire in the course of an active research career are major determinants of his or her professional versatility. How diverse are these skills, and how are they learnt?

Scientists whose research interests are highly diversified (§1.5), or whose research trails have taken them some distance across the scientific landscape, have obviously been forced to pick up many different ideas and techniques, like the one who said:

' . . . you can look at me as a sort of biochemist, a geneticist, a biologist. I work in all those sort of spheres − as well as, at times, a chemist. I mean I am both an analytical and genetic chemist at times in the laboratory, which I suppose really all good biochemists are. I am also an enzymologist, a classical biochemist if you like. I have done a lot of pure biological research . . . fish breeding, pure genetics − it covers a very wide range'.

Very few scientists actually 'migrate' abruptly to quite foreign fields of research (§1.6), but they often 'drift' a long way in the course of the years (§1.7). A specialist in crop protection, for example, who had moved from his original base as a mycologist, through plant physiology and biochemistry, into the hydrodynamics of spraying machinery, had crossed the frontiers of several scientific 'language areas' (§4.2). An applied physicist who started off on the design of an imaging device had crossed these frontiers in the opposite direction when he eventually found himself involved in research on the physiology of vision. Such career paths are not at all uncommon even amongst academic scientists − for example, the one who described his theme as 'the response of organisms under stress' had had to acquire many new practical skills as he drifted across biology from parasitology to pharmacology.

But what about the typical scientist who 'persists' in the same narrow research field for several decades (§1.4)? The stereotype is of the 'subject specialist' whose knowledge and skills are similarly highly specialized, being limited to no more than is absolutely necessary to get on with research in his or her chosen field. Many academic scientists certainly insist that this is how they define their 'interests' and their skills. This was indicated in the discussions by occasional remarks such as those of the marine biologist who would sharply distinguish between 'a fish man' and 'a crustacean man':

'And to some people they are totally different worlds. They really are different worlds, and they need rather different methodology, very different methodology, different thinking about it.'

Yet this point of view was not as common in the discussions as one might have expected — perhaps because, as was noted in the Introduction, the discussion groups were set up in a way that was not fully representative of scientists of this kind. In any case, it was quite clear that the stereotype is misleading in one very important respect: a scientist who has done research for many years on a very narrow *subject* would often be using very diverse *skills* in that research.

These diverse skills are needed because the scientific landscape is much more complicated than it appears to be on any scheme of classification by 'subjects' — and is always changing (§1.2). It is not enough to be familiar with, say, the neighbouring specialties on the contemporary 'map', since a new concept or technique may suddenly open up a connection with a quite different area of knowledge. Consider, for example, the agricultural researcher whose subject is the chemical treatment of aphid-borne diseases of cereals. This specialty stands, so to speak, at the junction of many different corridors of knowledge and technique — insect behaviour and physiology, plant virology, biochemistry, farming practice, and so on. A clue or a method for the solution of a current research problem might come along a corridor from any one of these areas. All the relevant skills in all these fields cannot be mastered in advance, but a few of them may well have to be learnt on the spot, if the occasion demands, in order to make progress with the research.

This point comes up again and again in the discussions:

'I think there are two ways to make advances in a problem. On is to know the whole thing so well, and the other one is crossing from different disciplines. And in that instance, you can't get very far (if you *are* going to cross) unless you know several regions very well. I think it's particularly important now in the biological sciences . . . .'

A scientist in another field suggested a reason for this need for diversity:

'It's a lot to do with the recognition of the degree of sophistication that's now involved . . . in solving the problems. The simple ones are no longer there. You've got to move in, you've got to bring in area that you're not expert in. You've got to look around and pool people together to solve this problem.'

In another research unit, where they said they had 'an understanding of a lot of different subjects', although this had 'all been directed down a fairly narrow channel', two scientists agreed on the same point:

A. 'I don't feel in a narrow little channel at all working here. It's helped me to broaden my horizons [so that] I could move out of this unit much more easily now than I would have done ten years ago.'

B. 'Yes, I will take that up, because I believe that in applying this rather special form of data for information you have to start to understand very diverse subjects and multidisciplinary projects to be really part of it. So in that sense you are getting a wealth of knowledge all the time by applying your science.'

It is now generally accepted that the *technological* expert in industry has to be a jack of all trades (§6.4). For example (Thomason 1970), an 'aluminium expert' ought not only to have a basic knowledge of the physics, mathematics and chemistry relevant to metal physics and metal chemistry, as applied to the various techniques of metal processing, such as forming, machining, cleaning, corrosion protection, etc. He or she ought also to be familiar with the testing, distribution, packing, storing and shipping of the manufactured product, and its uses in building, chemical industry, vehicle construction, etc. The striking fact that emerges from the discussions is that even a research specialist in *basic* science comes to need skills and knowledge which could not have been anticipated in his or her formal education in a conventional academic discipline

The growing points of modern science (§5.7), the problem areas of modern R&D organizations, however narrowly specialized in conception, are usually *multidisciplinary*. In due course they may be institutionalized into recognized 'specialties', each with its characteristic 'interdisciplinary' area on the scientific map (§1.2). But until that happens, those who research on such problems must be competent to bring to bear on them whatever skills and information they can acquire, by whatever means.

Thus, in many major fields of research, the distinctions between the academic disciplines simply do not apply:

'I regard my [more specialized subject within a specific subdiscipline of the earth sciences] as multidisciplinary. My way into it is through geology. I think people coming into the field with mathematics, or physics, or chemistry as their background are every bit as welcome in it as a geologist . . . . I learnt absolutely nothing, not one minute, about [it as a geologist]. That has been put right. Students now graduating in the first degree have more knowledge, but that does not help me very much because we are not really encouraged to recruit new graduates in the system . . . .'

Again, a scientist working on a particular class of materials had specialized in mineralogy 'which is not too dissimilar . . . in that [the science of these materials] is a combination of many disciplines'. A chemical engineer remarked that 'chemical engineering is rather a peculiar subject: it embraces a large number of disciplines', before describing the diverse skills that he had had to apply in various very specific R&D projects. Similarly, an expert on [−] systems referred to the 'huge variety of disciplines' that made his work so interesting.

In practice, of course, no one individual can be expected to be competent in all this multiplicity of skills. Few scientists nowadays work entirely on their own, and can turn to immediate colleagues for help or advice. Most R&D establishments are, indeed, staffed and organized along multidisciplinary lines in order to bring a variety of skills to bear upon their problems. In R&D projects of great urgency and extent (§2.3), the work is usually undertaken by a whole team of scientists, each of whom is made responsible for a different aspect. A lot then depends on the way in which teams are built up and managed (§7.4). If this labour is also deliberately divided along disciplinary or methodological lines, team research obviously allows a great deal of scope for narrow specialization of expertise. But this does not invalidate the general point established above. For example, membership of a project team forces experts in different subjects to co-operate closely, and thus become fully acquainted with each others methods, 'language' and scientific point of view. Indeed, as Irvine & Martin (1981) argue in their study of the subsequent careers of postgraduate students of radio-astronomy:

> 'postgraduate students, who play an important research role in Big Science, often acquire in the course of their work not just a knowledge of a narrow area of science determined by their thesis topic, but also a number of specific skills (in area like computing and electronics, for example) as well as more diffuse abilities (most importantly, the ability to work efficiently as a member of a large research team, perhaps organizing, coordinating and supervising work by other people)'.

The members of a small research unit, are usually ready to share their special skills and knowledge:

> ' . . . what tends to happen is that each area of interest . . . generates questions which you then throw at the experts in that field, and they either respond rapidly, or over a period of time, depending on how much the pressure is'.

But it is a different matter to call in an 'outside' expert to deal with some unfamiliar aspect of a research problem:

> 'And so, when I bring my problems to [another unit where they specialize on these other subjects] . . . the conception of what the problem is is difficult to get over, to start with. Then the methods, the ways they look at my problems just lead them into . . . they are the sort of thing I was doing a few years ago, and lead them into . . . sort of following a line of thinking which is not going to be fruitful.'

Scientists thus often prefer to become their own experts, because of the interpersonal difficulties of extramural collaboration:

> 'I have to draw on other people's expertise because my subject encompasses a whole range of different techniques . . . which I know nothing about, so I have to draw on other people's expertise . . . .. Some people are very good and go out of the way to help you, and so on, but there are plenty of people, particularly in [−] who just don't really want to know. They are only interested in doing their own little pet projects, and if your money is there for them to bolster their pet projects they are quite happy. But they don't really want to get involved . . . .'

4.5 Learning a new skill

The technical versatility of most experienced scientists (§4.4) is not only a valuable characteristic in its own right: it points to a capacity to learn *further* skills if required to do so. Some scientists describe their careers as a continual process of learning:

> ' . . . we've got this great armoury of computer programmes which we built up over the years, for all sorts of work, all sorts of purposes, and on top of that, every year or two − perhaps once a year − there is a fairly major new area on [−] and [−] − these are the two major ones − which you've got to learn a new area of mathematics on, and might persuade somebody else to learn'.

A researcher in mid-career (who said he now doubted his ability to master a new technique!) had:

> ' . . . done electron microscopy . . . various techniques in neuro-anatomy . . . that I have learnt by collaboration with a number of laboratories, and various kinds of techniques with electrodes which are required, and also using the computer . . . . And all of those have really been, I think, decisions because I felt in order to proceed that was the next thing to be done'.

For others, the learning experience had been more episodic:

> 'When you do shift from [one crop system to another] you face new problems. You've got to familiarize yourselves with a completely new farming area. One was very specialized and expert, of course, in a crop like [−] . . . and farmers turn to one in that respect, but now you're moving over into completely foreign territory.'

How do scientists go about learning new skills? One thing is quite clear: once they have left university, whether with a first degree or a PhD, they seldom go through any further courses of formal education. Some major technological establishments

have introductory training courses for their new recruits, and occasionally a young
graduate, after two or three years in a research job, will take a postgraduate course
to become qualified to work in another specialty. There are also a few cases of
people who entered the research environment straight from school, as laboratory
assistants or technicians, and who managed to get a formal scientific education
– even up to a PhD – by the heroic effort of studying part-time over many years.
But by the time a scientist is well established in a research career, the most that
he or she is likely to get by way of formal retraining is, say:

> 'an introductory course [in computing] at the Polytechnic . . . it was one
> evening a week for two months and after that it was just a matter of picking
> the rest up from a manual. But it was just the introduction, getting over the
> psychological barrier of actually talking to a computer'.

The overwhelming emphasis is on the informality of the process of learning and
on the extent to which it relied on personal initiative. Often it was a relatively
painless process of 'having to pick up a bit of geology as I go along', or of coming
into an establishment with 'purely an academic background' and 'having to learn
an awful lot more . . . on the applied side'. But what might have been lightly
described as having to 'sap the brains and bone up on [–]' in order to 'retrain
myself' so as to 'get involved in [a subject] which I knew nothing about' (this is
a pastiche of typical phrases) could mean teaching oneself the best part of a whole
academic subdiscipline (§1.1), up to the research front, in the course of a few
months, without formal instruction.

Indeed, many of them preferred this DIY approach. Had they felt the need of
expert tuition? No, it had to be done:

> ' . . . mainly through your own effort. I think that the best way of learning
> is trying to understand it your own self. It may be the least efficient way of
> trying to learn something, but it is to me, anyway, personally the best way,
> because the lessons tend to stick better if you learn things yourself. Whereas,
> if you hear and understand your opinions expressed by other people, and get
> something elucidated by other people, then it's less satisfying – but it's more
> efficient, perhaps, in some way'.

They did not underrate the magnitude of this task, but perhaps they saw that formal
instruction would not be very helpful in a situation where:

> ' . . . the problem is not so much one of thinking, but one of trying to
> acclimatize yourself to the terminologies of the new field . . . . There's a
> lot of jargon, and you've got to come to terms with it, and also appreciate
> the origins of certain concepts in a field which you are not familiar with
> originally. And certain terms and concepts may not be relevant to what you
> are trying to do in a new field, and trying to understand that – you know,
> the *non*-significance of certain concepts – is the key thing'.

In essence it was very much a matter of 'learning on the job'. There might be lot of background material to read up in books and articles, and one might go to 'university departments that were working in allied areas' to find out what was really going on. But that was not sufficient:

'I don't think you gain anything just by doing that. Because you have your own ideas; you start off; you make your own mistakes.'

In other words:

'It's a matter of working from a firm basis, isn't it? If you want to go and ask somebody, you should go to someone who has your own problems, so you can go and ask them to help you with your problems, and vice versa. But you cannot go blindly and say "you're [an expert in a certain field]: come and tell me how to do it".'

What this requires, of course, is the self-confidence to begin work in a field whilst one is still very ignorant about it. One may find oneself in the situation described by one scientist:

'. . . when I first went out to talk to a crowd of unknown people in the [–] division about problems I did not understand at all . . . . In the areas where I had been working before then, I always knew a few ways into the problem, to try. When I went to talk to the people in [–] for the first time, I had no idea what sort of approach to take.'

Or it could mean that:

'. . . I was doing things which, if you'd asked them, they would have said you need a man of great expertise to do this, but I managed to do it without the expertise. Ha! Ha! Ha! And somehow I managed to make things that worked'.

In other words, they had to break through the specialist's belief that:

'. . . the other man's job is incredibly difficult, and I can't think how on earth I'd ever be able to do it, but the job that I've been in a couple of years myself is incredibly simple, and I can't see why everybody else can't do it as well'.

Of course there are limits to the amount of expertise that can acquired in a short time in this way. In some R&D organizations, people may simply not stay long enough on the same project:

'We tend to move around fairly often . . . – perhaps too often to become the world's experts.'

In other cases, they come against 'language' barriers (§4.2), and recognize that the problems to be solved are 'beyond their expertise'. Nevertheless, these informal methods of 'learning on the job' are thought to be quite adequate for a scientist who has to 'pick up' a new subject or technical skill in the course of his or her professional career.

4.6   Teaching oneself

Most scientists nowadays, in almost every field of research, have to be quite expert in the use of computers. Yet those in their 40s or older can have had little occasion or opportunity to go to through a course of computer science when they were students. The way in which they acquired this skill is a good example of the process of informal self-education described in the previous section.

As many of them remarked, this was an area of technique where the pressure of change was felt particularly acutely:

' . . . we, of course, are not trained – all of us – as computer scientists. Increasingly, computers are moving in on the implementations side [of our research] – not only for our analysis, for our research results, which is well established, but in their actual translation into practice. All this sort of area is coming fast, and the technology is advancing extraordinarily fast, so that even the computer specialists find it difficult to keep up with. Now this is an area where we, in terms of sheer technology, lack the ability to move into, at the moment. And we are seeking ways to improve this, to develop this much more effectively'.

The challenge was seen to be one that could not be declined:

' . . . as scientists [we] are in a competitive field. [If] our competitors use these machines routinely, then it becomes the accepted norm of the subject that this is how it is done, and the work will become unacceptable if we don't reach the same standards'.

Nor could computing be regarded as a technique, that could be handed over to specialists:

'It's not just a new instrument: it's a new way of thinking about things. It's also a new way of dealing with experimental problems and theoretical problems . . . . I think it's an important part of our scientific literacy that we should be able to have contact with computers, because it opens up a whole range of possibilities that don't exist otherwise. Particularly if you have to go through an intermediary to the computer, then you miss more than half the possibilities.'

Even though many of the scientists faced with this challenge say they had 'not been mathematically inclined', or were 'not aware that I had any particular mathematical bent', or 'don't feel I had a mathematical flair' they found, in the end, that they could master this skill. But, as remarked above (§4.5), they would only take a brief course of formal instruction, before plunging in, like a child with a new toy:

'Well, to some extent it's like an acquisition of any skill: you need to be able to sit and play with it.[But you may not] have the facilities to do any [because] there is pressure of time on the machine. The last thing they want is somebody sitting there all morning with a manual, playing away. In our case, we have one in the laboratory, and that is the only reason we've come to use it — simply having it there every day to use. I think if I had to go along one hour a week to have done it, I would have found that very difficult, and if I felt that people were queuing up behind me I would have found that very harassing. If you are starting something new, it's like learning how to use any item of equipment: you need to be able to do it yourself, and have help and advice from specialists.'

What many of them found, when they had more or less mastered the craft aspects of computing, was that they had acquired an *intellectual* technique with much wider applications than they had realized. A lot of the numerical modelling techniques that they had originally learnt to deal with certain advanced engineering problems, for example, turned out to be just what was needed for a major project in geophysics, or for the analysis of a particularly complex industrial manufacturing process. This diversity of applicability is characteristic of many scientific techniques which seem very narrow and highly specialized in themselves: those who become really expert in them can move, so to speak, in tunnels beneath the scientific landscape, coming to the surface, like moles, in the most unlikely places.

4.7 Tacit skills

Scientific knowledge is so well defined and systematic that we tend to think that the expertise needed to produce that knowledge must be equally well defined. We are told, for example, that knowledge about fish is *ichthyology*, and is generated by *ichthyologists*, each of whom is a subject specialist (§1.1) on some particular genus of fish, or on the fish of a particular habitat, or some particular aspect of fish physiology, etc., and also commands a certain number of technical skills, such as catching, keeping and breeding them, or dissecting them under a microscope, or analysing their population data on a computer. As we have seen in this chapter, the knowledge and skills actually required for specialized research are often very diverse. It is important to remember that they also include an indefinable *tacit* element, which is often overlooked.

Following the work of Polanyi (1958), philosophers of science acknowledge the vital importance of tacit knowledge — intuitive, uncodified 'knowhow' — in the performance of specific technical skills, such as manipulating an electron microscope, or solving a particular type of mathematical problem. Such knowledge is of the essence of real skill (§1.1), and is only to be gained by years of experience. The geologist, for example, who has spent 10 or 20 years surveying the rocks of a particular geographical region will have acquired an invaluable stock of knowledge that could never be read directly off a geological map. This sort of knowledge is not merely the prime personal asset of the subject or technical specialist (see §5.2): it is a key factor in the division of labour in research.

But are the tacit elements of skills as specific as these examples would suggest? Having learnt to drive a particular car, for example, we discover that we have the essential 'knowhow' to drive almost any road vehicle. The geologist who has learnt to understand the detailed structure of one particular region will know how to get to grips with the geology of any other region. In one of the discussions a biologist emphasized his 'gut feeling about forests', which was 'indefinable but useful', and which indicated 'the intuitive feeling and interest you have for a particular subject that makes you content to work on it anywhere': this tacit understanding evidently applied to the subject matter of something like a whole scientific 'discipline', not just to a narrow 'field' of research. An experimental physicist indicated the narrowness of the tacit component of his skill by insisting that he 'would not feel competent' to take up work involving microwaves, because he had 'played around with microwaves in the past, and found them intractable things' − and yet would refer to 'a certain way of thinking' that characterized all the work in the large establishment where he had spent most of his career. Professional experience in a narrow specialty gives most scientists tacit knowledge that is applicable throughout the whole 'language area' (§4.2) in which they happen to have worked.

Indeed, they sometimes suggest that the most important thing they have learnt is 'how to do research'. This is thought of as a general skill:

' . . . I think that we underestimate the scientific method − the research method which is applicable to any subject. There are research methods which we tend to apply by instinct or experience [but] tend not to be trained in. In a sense, the nearest word to it is systems analysis − that approach − the systematic analysis of a problem . . . . I don't think we are paying enough attention to the methods of systematic analysis of the problem'.

To some extent, this is attributed to their education and early research training, where 'you learn to plan and organize yourself' and 'you learn to solve problems of all sorts'. Irvine & Martin (1981) report that over 90 per cent of radio-astronomy PhDs say that their research training gave them the 'capacity to undertake original scientific work' and 'ability to tackle and overcome complex problems'. The need for such training is emphasized by Bucher & Stelling (1977), who point out that initially students have no concept of how their research would end, when to finish a series of experiments, and so on, but would do their research one step at a time, without having a notion of a complete investigation. Similarly, Reuter *et al.* (1978) note the difference between 'apprentices', who tend to undertake unambitious research projects whose results they do not foresee, and more experienced 'professionals', who anticipate their results to some extent, even in quite ambitious projects, because they know how to apply reliable methods. Thus, the tacit knowledge acquired whilst doing a PhD is extended by further experience – for example, by learning the necessity of 'asking the right question',or even 'refining and finding the question' before beginning research.

It is not for us here to debate whether there really is a general 'method' for scientific research, transcending the methodologies of particular sciences. The significant point is that one of the elements of 'knowing how to do research' is knowing how to move into an unfamiliar field and get up enough skill and understanding to make some progress in it. Every professional researcher has had this experience, and has learnt certain practical strategies for dealing with it.

Usually, there is no time to go off and spend six months or a year in some university department or industrial laboratory to learn about the subject systematically. Quite apart from all the practical difficulties of arranging this, in a fast moving field 'you'd learn all the wrong things'.

Professional researchers know that they must either hire somebody who is already an expert on the subject – which may well be the right thing to do – or else work on it in relative isolation, as individuals or as a small team.

What they do know is that they must not get stuck in mere study. There may be a lot of reading to do, but it is vitally important to start up some actual research, however naive, in order to find out *for oneself* the nature of the problems in the field. If the problem one is tackling is very practical one may even come to the conclusion that it is best to 'put published literature to one side', with the knowledge that one can always go back to it for specific items of information.

The experienced researcher also knows that:

'if you start off without knowing anybody [it is difficult]: as soon as you know somebody then it all becomes easy. With hindsight, you wonder why on earth you didn't do it two years ago. [But] two years ago you did not know all these people, and didn't know how to go and ask. Just looking through . . . lists of people . . . [in] libraries . . . does not really say enough about what they are doing that is any use to you'.

In any case, one would know that one would have to have something of ones own to offer, in return for such advice:

'I would not dream of going through a list of people, and isolating somebody whom I thought would know everything about it, and walking along and saying "I am going to do the same sort of work that you are doing, that I know nothing about", because I would know the sort of answer I would give somebody who said that to me. I would say, "Well, am I not doing it – and not you?".'

One would thus be quite prepared for a frustrating period when nothing seems to happen:

> ' . . . it is very difficult to avoid the first year of mucking about. If you come totally fresh into a new subject that nobody else is doing, . . . it's almost inevitable there will be a year where you are just mucking about and playing around. But, on the other hand, . . . you learn a lot over that year, and it's sometimes a great advantage over someone who is long established in that field and has set ways of doing things . . . the first year is very valuable [although] it can be very frustrating'.

Experience of such periods of frustration in the past — and the recollection that they have usually been followed by genuine success in getting useful results — can contribute greatly to the confidence with which a scientist may enter a new field of research. As we shall see in §8.5, this sort of tacit versatility is a very important factor in the adaptation of scientists to change.

4.8   The personal factor

What we have seen in this chapter is that scientists are usually much more versatile — that is, they have a much wider range of relevant skills — than the narrowness of their research specialties would seem to indicate. They have often had to learn quite a bit about a number of other subjects, and have had to become competent in a variety of techniques, simply to do what needs to be done in a particular field. Along the way, they have become accustomed to teaching themselves new things, which is such a vital factor in the craft of research.

But versatility, in this sense, is no more than a *potentiality* for change. Many other factors determine whether a person *decides* — or is, at least, *willing* — to enter into an overt process of change, and eventually *adapts* to its circumstances. In the following chapters we shall look at the *motives* that might enter into such a decision, within the context of a professional career. There still remain those indefinable characteristics of 'personality' (§4.1) by which people differ so greatly, one from another, in their response to the circumstances of their lives.

Indeed, there is a common opinion that 'adaptability' is an innate characteristic, with a fixed distribution in any population:

> ' . . . [It's] the same in an ecological organization as in the Ministry of Defence — or in Roman times, when a charioteer had to decide whether he was going to go out and become a swordsman in the arena . . . . [A social scientist] could assess the proportion of people [in a research organization] who were unwilling to change, those who were willing to change, and those who were very flexible . . . . It is largely a matter of personality, certainly, when it comes to making oneself available to be adaptable . . . . Some

people will and some people won't . . . . The next stage is finding out whether you *are* adaptable. I believe that half the people are willing to have a go at it, and that some of them will find that they are not as productive as they hoped they might be.'

Another widely-held opinion is that people become inherently less adaptable as they grow older. This may well be the case, but is it because they are less *versatile* or because they are not so well motivated to accept new circumstances? There is little evidence in the discussion material to suggest that there is a general decline in research capacity through the middle years of a scientific career. Some people report that 'they find it more difficult to learn' new subjects, although 'you get there in the end, but it just takes you about three times as long as a new graduate'. This would apply, naturally, to advanced mathematics and to any field requiring a great deal of mechanical memory work. A rather young research group were scornful of another group who were 'older than we are, and had to retrain at a late stage in their career.' But cases were also reported of people who had 'adapted very successfully [at 58] . . . in the tail end of his career', and it was recognized that:

' . . . other qualities are coming in – experience, more of a commonsense viewpoint. You don't rush in, waste a lot of time testing everything that comes along. You can size things up. This often compensates'.

Quite a lot of research has been done to test the hypothesis that there is a decline in research competence with age, but with similarly equivocal results. For example, Gieryn (1979) observed a number of minor changes in the patterns of problem choice of older astronomers, but quotes Reskin (1977):

'Extensive, albeit methodologically inadequate, study has failed to turn up any convincing evidence of a strong simple relation between age and the productivity of individual scientists. This relationship still awaits theoretically informed and methodologically rigorous investigation.'

Fox (1983) comes to much the same conclusion. We might as well accept the opinion expressed in one of the discussions:

'I don't believe, on the whole, that people go off with age as much as is said. I think you've got bad scientists who start off not very good, and then when they get to their 40s and 50s they pack up altogether – and we all know a few! But I think the good scientists, most of them anyway, keep going very well, and I know many in their 60s who are still pretty smart, and in a way you're sorry when you see them pack up . . . . On the whole, they're the better ones to start with – and I know very few . . . . I know of very few who were damn good scientists when they were young who haven't stayed damn good scientists.'

Indeed, if we were to accept the notion that adaptability is completely determined by innate 'temperament' and/or age, then the present study would be futile, for there would seem to be little that anyone could do, before or after a change of circumstances, to help any particular person to adapt to it. My own feeling is that the evidence presented here is against such a strongly determinist principle, but all our everyday experience of people suggests that there is a good deal of truth in the precepts that people do get less flexible as they grow older, and that no amount of training, or wise counsel, or managerial finesse will affect some of the patterns of personality 'established when we were kids'.

Research itself is certainly not a homogeneous occupation, calling for one particular type of ability, or one particular type of personality. It is true that most scientists still reproduce the traditional stereotype of the 'ivory tower people who would never change at all', 'persisting' all their lives in a narrow career path (§1.4). Nowadays they usually work together in large teams on relatively 'urgent' and 'extensive' projects (§2.3) where there is usually adequate scope for the division of labour into narrowly specialized work roles, so there are still plenty of career niches for the introverted, obsessive specialist who finds psychological security in becoming the world expert on one little subject.

But there are also places for those with sloppier, broader, more extraverted abilities and temperament. They are needed to 'go round the perimeter and collate problems — interpret problems in a more general way'. In industrial R&D organizations they respond to the increased effort to 'involve people . . . in the actual problems of the process' and to 'establish and maintain contact with production'. A researcher working on a large practical problem notes the contributions made by various experts, but makes it clear that he had himself made a significant contribution, even though he had only 'delved into lots of topics', but was 'not an expert in any one'. In the interviews it was noticeable how often a scientist would deprecate his or her capabilities as those of a 'jack of all trades' (§4.4), whilst indicating that it was their skill as a general 'problem solver' that was valued by their employers (§9.3).

Nevertheless, within this diversity, all scientists stress 'the essential difference in temperament between scientists and non-scientists, problem solvers and non-problem solvers'. We have already remarked (§3.4) on the notion of personal commitment, of being 'self-winding', which is the stereotype that scientists have of themselves and of each other. This is absolutely fundamental to the theme of this chapter:

 ' . . . It's just a matter of being positive, saying 'this is interesting'. It's not
 a matter of whether you can do, or you can't do it . . . . If it is interesting,
 you can teach yourself anything . . . . Being a scientist forces you to be
 versatile, because if you are not versatile you will hardly solve anything.

Because you keep running against problems of technique, problems of knowledge that you haven't been aware of before, and unless you actually fight your way through them you will never actually produce anything, or get anywhere . . . . The very nature of it forces you to be versatile.'

Many of the scientists in the discussions stressed the psychological importance to them of 'doing things for themselves', of 'not wanting somebody else determining my career', of 'being able to do the research that you want to do', of having 'complete academic freedom', and thus being able to 'think independently' and 'operate autonomously' (§9.3). They refer to the belief that:

' . . . in general it's the people who have been against the consensus, who have been in the real innovative research areas, who are responsible for the great leaps forward in science'.

and they naturally prefer an organizational environment where:

' . . . it's a "bottom-up" kind of initiation of work, on the whole . . . within the areas of responsibility of [informal] "cells" [of expertise], rather than "top down", when people say "well, we want you to look into research on a particular application".'

Few scientists nowadays are actually given the freedom and independence to which they thus attach such value. Indeed, this may be, as Eiduson (1973) remarks:

'an overdetermined reaction, because the scientist feels that he knows and can trust his own personal resources, but he has doubts about relying on others and trusting them'.

But even if it is a neurotic symptom or a myth, this commitment to the ideal of self-motivation is the essential instrument by which scientists carry themselves — and can be carried — through radical and abrupt changes in their careers.

5

Motivation

5.1 Rationality in career decisions

There is all the difference in the world between what one is *competent* to do and what one *chooses* to do. A technically versatile scientist may not take to an abrupt change of field for the perfectly good reason that there does not seem to be much in it personally for him or her, however desirable it may be from an organizational point of view. Research scientists are not slaves or soldiers: there is little point in trying to force them to do jobs for which they have absolutely no aptitude or taste. When they say that a particular change of field was 'involuntary' (§4.1) they usually mean that they were not able to continue in the work that they were then doing, not that they were given no choice at all in what they did next. In fact, most R&D establishments try to offer their redundant scientific personnel a 'reasonable' choice of another job within the organization, rather than giving them the sack, and even those who are unemployed are allowed to exercise some preference in the type of work they seek (§9.1).

The factors that enter into such a choice are inevitably highly personal. A person in such a situation is bound to respond to feelings and tastes that they can scarcely state in words, let alone justify to other people. But they would also take into account a number of practical considerations that can be formulated fairly clearly and discussed with friends, colleagues and employers. Some of these considerations, such as the need to stay near an ageing parent or reluctance to work on a military project, are too individual and private to be set out and analysed in general terms. Our concern here is primarily with a limited part of the whole rationale of decision in such cases – the part that refers to the effect of the change on a person's professional social role.

One would not expect scientists to rule their lives solely by rational calculations of their selfish interests in the performance of those roles. Indeed, they often insist (§7.1) that they 'never actually plan their careers', and that they 'have never thought of, at most, two years ahead'. Nevertheless, when faced with the need to make a hard decision about their future work, they are bound to give such considerations high priority. In other words, *career motives* play a major part in the initiation of and adaptation to any major change of subject, or employment. These motives are particularly complex for scientists, because of the interdependence of reputational and organizational factors (§3.3). It is not just a matter, on the one hand, of deciding whether a particular course of action is likely to enhance ones standing in the scientific world, or, on the other, whether it will be an opportunity to show ones value as a loyal servant of the company. Sometimes these factors work together, so that it is felt that a move with promising reputational implications must surely lead to organizational advancement. In other cases, there may be a difficult choice between, say, trying to stay in basic research with the hope of promotion up the 'scientific' ladder (§7.6), and working on a technological project which is most likely to lead into a job on the production side of the organization (§6.4). This is a peculiar feature of R&D organizations that is scarcely considered in the literature on personnel management in general.

Nevertheless, it is quite clear that the only way to sustain a reputational career is to keep going as a researcher in the same, or another specialty. But there is often the alternative of taking up some other occupational role, in the management or administration of an R&D organization (§6.2), in more general managerial or administrative work (§6.3) in the provision of technical services (§6.4), or in education (§6.5). Some of the considerations that might motivate or inhibit such a change of career are discussed in Chapter 6.

5.2   The value of expertise

The division of the labour of research into minute areas of specialization (§1.2) makes expertise in a particular area into a rare commodity. Anyone who has gone to the trouble of acquiring it has very good reason to make the most of it by continuing to put it to use. All the arguments for the necessity, possibility or desirability of moving out of a research specialty must be set against this elementary fact.

The intrinsic value of individual expertise is so central to the scientific enterprise that it scarcely needed to be debated in the discussions. The conversation would revolve, rather, around strategies of acquiring or keeping the expertise that was needed for a particular programme of research. Suppose, for example, a research group was moving into an unfamiliar field: was it:

> ' . . . efficient to move people into a different field of that sort, and start
> from scratch, or . . . better to bring people in from outside who already have
> experience . . . . The programme might have gained a year or 18 months
> by bringing in people from outside . . . . You can actually get a much better
> job done *cheaper* . . . . It may be better to pay a professional to do the job

for you, rather than expect someone to try and pick up the reins . . . . If you are moving into something [quite different from what you have been doing] there is no way you are suddenly going to pick up all the information'.

Even when the research programme is not changing rapidly:

'There is so much a need for the steady job of building up expertise. There is nothing obviously coming out of that until a problem actually arises where that expertise suddenly is absolutely vital. Somehow, as a research manager, you have to be able to keep that underground [basic] research going along in your department while you are making your fast contributions on [applications].'

There can be no denying the importance of real professionalism:

' . . . there's a distinction between some people [who can] just write a program, and people who can actually sit down and design . . . a program in a sensible way so that it doesn't run out of machine time . . . . You've got to have somebody who knows what they're about, and produces something sensible'.

The 'do it yourself' impulse of the self-winding scientist is always up against the reality that:

' . . . if you've got a few results you want to analyse . . . it's probably a lot more economical in terms of man-hours etc. just to get [−], who's got the experience, just to run a programme through, and get the results, than actually sit down and spend weeks trying to convert it'.

In every field:

' . . . there is a . . . reservoir of knowledge that is built up about this subject over years of research, and that is something we cannot pick up in a year . . . quite a lot of the technology [in a novel subject] could be picked up, but not the fund of knowledge: who is working in the subject, what has been discovered, where to turn for information − that is something you acquire much more slowly'.

In other words, the work of research depends on having access to the expertise gained by many years of experience. Many scientists feel themselves to be in a position where they can say:

'If I was changed tomorrow, I don't think the work [that I do] would continue. There is nobody else in the group doing the work.'

It may have come as a shock to them originally when 'having been here only a few months, I suddenly found out I was the expert in this field, or nominated as the expert', but they soon fall into this role, and discover that they have a heavy personal investment in being one of the few people in the country, if not in the world, with that highly specialized expertise.

This sense of having an investment in a particular role in a particular establishment is enhanced for scientists who work with expensive, highly specialized equipment. If they were to move to another establishment, would they be able to get the apparatus they would need to continue their research? If they were to move to another field, would they not need similarly expensive apparatus to get started?

> 'Very often, of course, you can go to a going concern, and it will have resources. But sometimes the issue of resources and what we can do with them, or whether you can get access to them is a very crucial element, especially if you are trying to start something from scratch.'

The equipment they are now using, which they may have built up by their own efforts over a period of years, is itself part of their expertise.

Specialized scientific expertise, whether intellectual or instrumental, is essentially a wasting asset, which must be kept continually in repair. It is not an investment to be relied on, for it is always at risk of being made obsolete by scientific or technological change (§1.2). Nevertheless, the high valuation that scientists place upon it, in themselves and in their colleagues, is often fully justified, and they have very good reason indeed to reject any move that threatened to rob them of their personal stock-in-trade.

5.3   Building up a reputation

Some scientists are 'foxes' (Gieryn 1979), whose research interests are so diversified (§1.5) that they do not claim to be real experts on any one subject. They make a name for themselves by the quality of their contributions to a variety of disconnected subjects. But the majority (§1.4) follow the strategy of the 'hedgehog' in getting to know 'one thing of much importance'. Recognition of this mastery of a particular field by *other* scientists then becomes the mainstay of their career. Any change of field is thus a threat to the *reputation* they have laboriously accumulated over the years.

Scientific reputations are intangible and often ephemeral. In academic science (§2.1) they are customarily based on, and bolstered by, *publications* – that is, by primary research papers published in the open scientific literature. Scientists may be scornful of the cynical dictum 'publish or perish', but they recognize its validity – especially for other scientists in other countries:

> 'One gets the impression their whole career prospects depend on the amount of paper work they produce . . . . We had a chap here . . . who recently . . . went to work in the States, and this seemed to be his sole object in life, and this what he is told – that unless he publishes *x* number of papers in a certain time he will not have his job.'

Of course, everybody knows that sheer quantity of publications is not sufficient to establish a reputation: it is also necessary that other scientists should take them seriously:

'. . . your image as seen by others is a kind of relative evaluation of what you are doing, and I would make this distinction between what appears to be citations [i.e. references to your papers in other scientific papers], because . . . you say "citations in the literature" – but how many citations? It is one thing not to be cited and it's another thing to be ranked number three [in the world, for the total number of citations in a particular field]. There's an enormous range between catching on at a certain level and actually being out and out successful'.

Scientists are well aware, however, that scientific papers are not cited or otherwise recognized in proportion to their intrinsic quality. It is almost inevitable that extra attention is given to the work of a scientist who has already published a number of papers in a particular field, who has dutifully attended the relevant scientific conferences, and has otherwise become 'visible'. Gieryn (1978) notes that 'the publications of *young* scientists are more highly cited if they are concentrated in one or a few problem areas', whilst 'for older scientists' (who have already established a name) 'they are more highly cited if they are distributed among a large number of problem areas'. In other words, it takes much longer to become an *acknowledged* expert on a subject than it actually does to become reasonably competent to do research on it. Scientists who are concerned about their reputational careers are undoubtedly well advised to concentrate on a particular topic until they have achieved this standing in the eyes of their colleagues.

In fact, a scientist does not have to be outstandingly brilliant to get into a position to say:

'. . . I'm amongst a handful of experts in this particular field, and I can't imagine any other field where I wouldn't be way, way down the hierarchy.'

For this reason 'it must be very difficult to change your specialization if you have established a world reputation', and are 'recognized by some small fraternity worldwide that understands what you are doing – the value of it'.

This is not just a matter of ego gratification: there are material rewards at stake. In many R&D organizations, building up a scientific reputation is also the way to *promotion*. In many research council establishments, for example, this is still the official policy in relation to Individual Merit Promotion (§3.3, §7.6):

A. '. . . you get a reputation amongst colleagues and peers for specialization. If you don't specialize, I am afraid it, to some extent, affects your own personal development. For example, take the case of the DSc [degree], which many people acquire as a natural progression of their interest in a narrow field. It is considered that if you want to be an SPSO by "merit",

you've got to have a DSc. Now if you change your discipline, of course that
is just not on . . . . In certain areas of the hierarchy, if you are not recognized
as a specialist and an expert in one narrow discipline you are at a
disadvantage.'

B. 'Yes . . . . The evidence has been that people . . . who are highly
specialized have been able to get merit promotion, again to some extent
through the DSc system. In fact, not so very long ago, one of my Directors
more or less took the line: "[if you] get a DSc, that's it; we are willing to put
you in for an SPSO. Or, alternatively, if you get an SPSO, we will put you
in for a DSc". It was that sort of thing'.

In another establishment there was reference to a researcher who simply refused
to help people working in other fields, presumably because 'he was singling
himself out for special merit promotion'.

As we shall see (§7.6), most R&D organizations now have alternative sets of
criteria for promotion at the senior levels. But the early organizational careers
of many scientists are still thought to be at risk if they are 'not going to get papers
out'. It is only after reaching 'career grade' that 'the chances of getting up to the
next grade are much less [dependent] on the total number of papers, and therefore
you can take a broader perspective'. Thus, one of the major objections to a change
of research field is the inevitable delay of 'three or four years before one really
produces any publications' − a delay that might be tolerable for the leader of a
research group, but 'that's bound to reflect on the promotion prospects of the more
junior members'.

This is the practical aspect of one of the most important points in the present
study. Scientists are uneasily aware that the 'reputational' status they labour to
acquire in a particular field has to be kept under continual repair, and cannot be
transferred from field to field. A scientist moving to another field realizes that:

'. . . it's making a positive decision. I am not going to trade on the status
I have in [the area I am leaving]. I am going to cut myself off from that area,
and go into something completely different'.

Quite apart from its career consequences, this has serious implications for the individual psyche.:

'. . . if you go right back to scram, right back to square one, you've lost
everything. You've lost your authority, you've lost your responsibility.
Recognition is an important thing for scientists, who I think tend to be under-
paid for what they do, and so it's their own self-satisfaction that's important'.

Their whole upbringing has taught them that they must not make mistakes − and
yet:

'. . . when you move too far out of the field you are familiar with . . . you
may make a fool of yourself . . . ; You have to be extremely careful, if you

are going too far afield, . . . to make sure that you don't trip yourself up and get egg on your face'.

It is not as if one were moving into completely unknown territory. The attitude of the existing inhabitants must also be taken into account:

'[A scientist who] has been given a bit of a nudge to go [into a more applied field] is finding it a bit of a strain because . . . he has moved from a familiar area, where he is a bit of a king pin . . . to an area where at the moment he is a bit of a new boy . . . . People and scientists are quite jealous, and quite rightly so, of their . . . scientific reputations. This is where their . . . job satisfaction comes in, and if you find yourself moving into a new field, you are taking a risk, you are sticking your neck out.'

There may even be 'resentment from other scientists in the field' because another person coming into it is going to give rise to 'a bit more competition', and perhaps 'eat into funds'. A group of scientists contemplating such a move might well ask themselves:

' . . . how we would feel if somebody hustled into *our* territory. Would we be resentful? . . . If you regard that there is a need that you can't quite cover in a certain area . . . you welcome it . . . on the other hand, you obviously try and preserve elements of your work on which your continued reputation is going to be based'.

5.4 Projects and commitments

The work of research has its own rhythm. Even in an R&D organization engaged on relatively 'urgent' problems (§2.3), there must be reasonable continuity of personnel over related projects as they come along in succession. Frequent individual changes of job can disrupt this rhythm, as shown by the effect of the policy of some organizations (§6.2, §7.3), especially in the industrial sector, of deliberately moving 'very bright' people around — 'two years here, two years there' — in order to train them for higher management:

'That really does not get the research work done. That trains a few people, and can even give R&D a bad name . . . . In the end, you end up with the people supposedly doing R&D, and in fact all being trained, all the time.'

The very long projects that are characteristic of R&D in a discipline such as agriculture have a strong tendency to immobilise those who are involved in them:

' . . . most of the experiments that we are involved with are planned . . . to run over a course of up to ten years, and so you are looking for long-term effects . . . . You are putting an awful lot of potential ideas into the open, which you hope you are going to harvest later on, and so there is a sort of incentive then not to move'.

This tendency is not necessarily due to managerial policy; as we saw in §1.7, it comes from the researchers themselves:

'one has a commitment to the programme you are trying to see through, and the experiments you started years ago and you want to see through . . . . Whilst we mostly feel that we are making a contribution to your particular branch of science, then one tends to stay with it, and draw a lot of satisfaction and motivation from that'.

Thus, a scientist who is fully committed to the research projects he or she has in hand is bound to feel that:

'the only incentive at the moment for making some decision about [a change of field] is to say "right, this piece of work is finished: I can't see myself making much more contribution to it, therefore it's a good time to go".'

But how seldom this moment seems to come! One of the troubles is that even a narrowly-focussed researcher may be involved in several projects on closely related problems (§1.5), and these never seem to finish simultaneously. As soon as one is ended, another is begun, thus generating an endless rope of finite, inter-twining strands. A scientist who was moved after having been in the same job for seven years said that he had 'created a little bit of fuss about it', because:

' . . . it came at a very bad time with respect to the research that I'd been doing . . . . I'd got about five or six loose strings, which I just didn't have the chance to tie up before I left, so that there were various pieces, topics which I'd progressed three quarters of the way along the road towards solving . . . .

In other words, research is an open-ended activity, which *always* seems unfinished:

' . . . that's what annoys people. Generally there is something that's unfinished whenever you're moved. In my case, I was given more warning than most, but there were still things unfinished, and . . . although I accepted that I ought to move when I did, I still resented it to a certain degree'.

Indeed, one of the characteristic responses of scientists to enforced change of subject is a determined effort to 'keep my own field of research going as well', which might mean 'doing a rear-guard action', and 'trying to manoeuvre myself so that I could at least do something that was interesting to me'. What they feel is that:

' . . . once you've got interested in something, and provided you've given something to that research, it's very difficult to ignore it in the future. It's something you've become used to, and it's become part of your way of thinking about the subject. And then it's very difficult just to ignore it.'

The scientist who says that he would 'feel very sorry indeed' if he had to drop his original line of research justifies his 'interest' in it because 'there is no end to it, and I can still keep on going indefinitely, thinking about problems'. Even what is admitted to be a 'narrow research line' is said to offer 'many stimulations, taking those ideas, bouncing them off research colleagues, nationally and internationally'. A problem that 'really arose about a hundred years ago' and 'has not been cracked' has been the 'direct concern' of a scientist 'on and off' for some 20 years. Of course there may be periods when a longstanding problem should be dropped for a while:

'In order to make a lot of speed, it's very helpful to deshackle yourself very rapidly from some things. On the other hand, you will come back to a point in time where some of those original issues are actually rather important. So it turns out that they have got something to offer when you run out of breath. So that's one form of re-tooling − not to give up.'

But the pace of research is such that if a scientist leaves a field entirely for more than a few years, he or she would find it 'completely changed' if they came back, and there would be 'a lot of catching up to do'.

A positive feeling of personal commitment to a particular problem or research programme is evidently an extremely important consideration for a scientist facing a decision to change to another field. This feeling is not entirely rational in terms of worldly ambitions, but is perfectly understandable for anyone who has experienced the 'excitement' of research and discovery. It stands for the continuing hope of getting some return for the *psychic* investment that he or she has made in the field − the return that would surely come from the eventual solution of the problem, or the successful completion of the programme of research.

5.5 Fear and failure

Research is usually an absorbing, satisfying job (§3.4). There is normally plenty to do, and one works at one's own pace. It is scarcely surprising that:

' . . . as long as they are reasonably happy with the work they are doing, [scientists] will sit back until, from above, comes an instruction or invitation to apply for another job, or to move to another area . . . . Their immediate reaction is usually somewhat apprehensive. Secondly, they want to know what's in it for them . . . . Thirdly, if they feel that their manager really wants them to go . . . then they will go'.

The barrier to a move may well be 'purely a question of self-confidence . . . many people are worried about whether they can cope with it'. In other words, 'people tend to stick where they are safe'. As Sofer (1970) observes, people may even be 'reluctant to try their luck elsewhere, in case their current employers had made a correct estimate of their abilities'.

This element of anxiety in the rationalization of immobility is common to all organizations employing skilled personnel. Bailyn (1981) sums it up as follows:

'Movement is seen as beneficial by both management and professionals, but both also fear it. Individual employees fear it because of insecurity: they are not sure they will be able to learn the new technology; they fear that during a period of retooling they will lose their performance ranking and their salaries will suffer. They are afraid to risk movement for fear they will fail. Their present position, although they complain about stagnation, is at least known and secure. Management, too, is afraid of movement in the organization. Each individual supervisor fears the loss of the group's best people. And yet the advantages are also obvious. Managers observe the rejuvenation when professionals are put on completely new tasks; and internal movement allows them to manage better the effects of differential rates of growth in different parts of the organization.'

Deeper than fear of an uncertain future, and even less readily voiced, is the feeling scientists have that being moved is 'a slight on their abilities'. As Sofer (1970) points out:

'Movements within an organization always carry a freight of meaning (sometimes intended, sometimes not) and it is not always easy for the person to assess the meaning of any particular move for his long term aspirations.'

A very high degree of personal autonomy is one of the attractive characteristics of science as a vocation (§3.4), but it ties career progression closely to eventually getting some results. However skilfully performed, work that is left incomplete counts for nothing in the eyes of the world:

'We've had complete freedom and opportunity, and therefore any suggestion for change is a suggestion that we've failed in the area we've been working in.'

It is part of our normal comprehension of the hidden meanings of conversation to interpret a suggestion for change – especially from a superior – as a hint of failure. This interpretation is enhanced by the conventional hierarchy of esteem in the scientific community. When scientists say:

'We are highly paid scientists, so we have to do good science, rather than the kind of jobs that a test work establishment would do.'

They are not merely expressing the conventional prejudice in favour of 'science connected with very basic science, rather than applied science' (§2.3), for their notion of 'good science' certainly includes a strong component of practical relevance (§3.4).

They also feel that research that is undertaken by choice is somehow of higher quality than research done to order:

'People who are not . . . tied to short-term jobs, but tied to doing more basic research on a longer term, they − well, it seems to go with wanting to do that sort of work . . . . It was my view, when I was doing that sort of work, that I did not wish to change direction for somebody else's whim. I could see where I wanted to go, and I wanted to keep going in that direction . . . . People who might be suitable for doing that sort of work . . . probably have very strong ideas as to what they want to do anyway, because if they are good people − intellectually good − then they ought to know what they want to do, and they don't want to be told by a committee or by somebody else to "go and do this", and they certainly don't want to have the fear that if they say "yes", this time, that, in a year's time, the priorities around them change, and they will be told "go off and do that".'

Under these circumstances, the very idea of being 'moved' − rather than 'moving oneself' − is humiliating. At the heart of the matter is a deeply engrained, highly traditional idea of the nature of success as a scientist. This idea is seldom clearly articulated by scientists themselves, since it is essentially a personal expression of the ideology of science as an entirely autonomous profession (Barnes 1985).

It sometimes seems, indeed, that specialization itself is valued, in the same spirit, as a token of success. Perhaps this explains the experience of two scientists who had violated this tacit norm:

A. ' . . . having made the change you are tended to, in certain circles, to be considered something of a strange fellow'.

B. 'Oh, absolutely. I mean the probing and questioning I had from people I know . . . within the [−] system − generally friends and colleagues − . . . was quite incredible. They just could not understand how I had achieved it . . . and why, what was my motive in doing so. It was most odd.'

This attitude is not peculiar to science. In general, as Sofer (1970) remarks, 'a person who seeks to abandon what others accept as his identity holds himself up to ridicule'.

Ideological considerations such as these must be taken seriously by anyone concerned with any highly-qualified professional occupation, for, as Becker *et al.* (1961) point out:

'When laymen think about occupations with whose inner workings they are not familiar, they must rely on such ideas as have common currency. When they assess occupational success, for example, they use those criteria they are familiar with, criteria that can be applied to any kind of work, irrespective of its unique content, criteria like money, hours and prestige. But when a member of a particular occupation thinks about his own aspirations, he is likely to use, instead of these common cultural standards, an esoteric set of standards common only amongst his colleagues. For all occupational cultures contain, among other things, a set of standards for judging one's own work which can only be understood by those who know from their own experience what the problems of that kind of work are. The member tends to set his own standards and assess himself in relation to those standards which lie in the occupational culture.'

5.6 The dangers of persistence

The possible disadvantages of staying on in the same field of research are much less obvious than those of moving elsewhere (§5.5). Yet they can be very serious, as most scientists are uneasily aware. Although their immediate perspectives (Becker *et al.* 1961) are dominated by the expediency of going on with what they are doing, in the longer term they are often frightened of being stuck in one job too long.

This feeling, which Bailyn (1981) observed at all levels in American industrial R&D, is characteristic of all hierarchical professions. It is primarily a fear of *career stagnation*. To reach the top in a bureaucratic organization (§6.1), it is essential to rise through a number of ranks, usually by moving on from job to job, gaining experience and credit for doing them well. To stay for ten years or more in the same job — even with one or more steps of formal promotion — might well be interpreted as a symptom, if not actually a cause, of lack of professional success.

The scientist who remarks that he is 'at an interesting stage in career development', in that he 'can go no further', is thus very conscious that:

'. . . it's a fact of life . . . that most people reach their career potential [when] they have probably got in the order of 15 or 20 years to retirement'.

He would like to move to another job, if he could, to break this log jam in his organizational career (§3.3), if only by showing that he is indeed capable of bearing the responsibilities of another type of work.

Long continuance in a very narrow field of research work can also have serious consequences for the *reputational* career of a scientist. These consequences are not at all unfamiliar in the scientific world, and yet they are not always foreseen by those who suffer them. They therefore call for particular attention in the present study.

The most obvious danger is of a sudden 'breakthrough' that devalues expertise. Scientific progress is no respecter of persons. Radical cognitive or technical change (§1.2) may render obsolete, almost overnight, the specialized skills and detailed knowledge on which many researchers have been banking for their future careers (§5.2).

This risk may, however, be somewhat exaggerated. Major scientific revolutions are rare events which cannot be legislated for: scientific change normally takes place piecemeal, over periods of years or decades (§1.2). Active researchers can usually adapt to the decline of specialties and the growth of new ones by allowing their interests to 'drift' slowly across the scientific landscape (§1.7), and acquiring new skills along the way (§4.4, §4.5).

The real danger is that the obsolescence of their personal expertise can be so gradual that individual scientists fail to recognize that it is occurring. They are naturally continually worried by the effort needed just to 'keep up' with their subject:

' . . . the pressures that I felt are the ever-changing technologies surrounding the work I do . . . . I feel that I shall never learn what I should do to be competent in my job and to use what is available to me because of the acceleration of technology'.

A scientist in this situation may be tempted to specialize more and more, exercising existing skills within a narrower and narrower field, until (Reif & Strauss 1965):

' . . . he resign[s] himself to pursue his own specialty – even if it has become more remote from the main stream of his science – and become[s] content to watch younger people pursue the new areas'.

Such cases are very common in the scientific world:

'Objectively, looking on the outside, you can see people who [work] in fields which are being left behind. They have "gone stale", we say; their work is "not very interesting"; there are a number of phrases which are very familiar . . . .'

'I know people working in [–] who were doing absolutely first-class work, but [the assumptions on which their research was founded] have been broken now, and once that's broken that's left them a bit high and dry with regard to resources and students, and so on. I wouldn't necessarily say their work is not good – quite the contrary. That's different from the case [of] a unit that's gone off the boil . . . in an area which itself has not been left behind.'

In more formal language (Gieryn 1979):

'Scientists who continue to pursue problems with solutions apt to be of little interest will probably not receive much recognition, even if their substantial previous investments in the research make them more efficient than anyone else in finding the solutions.'

This is the syndrome aptly labelled by Chubin & Connolly (1982) as *undue persistence* along a research trail (§1.4). They explain it sociologically as a consequence of:

'positive feedback relationships between movement down a . . . research trail . . . and subsequent availability of support from the environment.

Success breeds success . . . . In terms of [an] incremental cost – benefit model, once launched on a research trail, a researcher or team is likely to persist in it well beyond the point at which additional studies can be justified by their scientific yield. If research costs decline rapidly after the first few studies, publication is valued for its own sake as well as for the size of the contribution reported, and scientific yield invariably reaches a point in the trail at which incremental yield becomes small, then undue persistence in worked-out, low-yielding research trails is the likely consequence'.

Chubin & Connolly go on to consider the conditions favouring this phenomenon : 'legitimacy' –, i.e. the existence of recognized research specialties, with established publication outlets, etc.; the system of funding by peer review of research proposals; access to existing research facilities in the locality; and the relationship between senior and junior researchers in postgraduate and post-doctoral research training (§3.3). These conditions relate particularly to basic science in the United States since the Second World War, but they apply to all research organized in the academic mode (§2.1). The essential feature is that they give rise to a 'Matthew Effect' (Merton 1968), which enhances the material resources available to researchers of established reputation – and helps them to increase their reputation still further.

This is a primary feature of the *communal* environment in which scientists individually pursue their reputational careers, and which they take as given in assessing the advantages of staying on in the same field (§5.2 – §5.5). This environment is not quite the same in all countries, and is changing slowly towards more 'collectivized' forms (§2.2). Thus, for example, the funding of basic research specialties is subject to much more utilitarian criteria (§2.5) than it used to be, so that an academic scientist can no longer rely on a reputation for doing 'good science' to win continued support. Nevertheless, declining research specialties (§1.2) also 'persist unduly'. As every research manager knows, they are very difficult to kill off by administrative action (Irvine & Martin 1982), and continue to attract resources and personnel long past the point where they are contributing significantly to the advancement of knowledge as a whole (§9.1).

This feature of the research environment is not usually so compelling for scientists working in formally structured R&D organizations (§2.2), as it is for those in academic institutions. Easily read signals of declining interest in a particular topic may come from higher up:

'It's very demoralizing when you see the priority associated with a task fizzling away, and people you are working with and dependent on for the success of the whole project filtered off into other tasks. One understands the situation, but if you are the unfortunate person who has worked on the topic for four years, and you happen to remain in that subject, it can be quite demoralizing.'

The enthusiasm of collaborators may be seen to wane:

'. . . I started to get a little bit frustrated . . . because I believed that the actual division handling the product [of my research] were deciding to sit back, instead of taking the initiative and deciding to screw the opposition right down. They said, ''well, we are all right now for a while: let's sit back and carry on'', and that's where, more or less, they still are now'.

General support may fall away for other reasons:

'[I felt] that I was certainly into the law of diminishing returns in the job that I was doing, partly because I had been in it so long, and partly because of the circumstances of the laboratory, which were contracting. You are having to chop off bits of work that you were interested in, and so on . . . . The department as it was does not now exist.'

But even in industrial and governmental R&D organizations 'there are people working . . . that are doing the same things that they were doing 15 years before'. They themselves may feel that 'they're quite happy in that little niche', but the people around them begin to notice that they are 'going stale'. It is not that they obviously neglect their job: at first they simply 'do this well as an experienced person', but show that they are 'not particularly interested in it' and are 'not going to think about it any more'. As the malady develops, the symptoms become more marked:

'They switch off. They switch off. And they do the minimum of work, and they look upon coming here as a chore, between 8.15 and 4.50. There are people who get into work that gradually winds down, and they can see no future for themselves. There are two choices: leave − well, it's a very pleasant place to live here − or wind down.'

The peculiar danger of 'undue persistence' is that it is so insidious:

'It's a very long process in some cases . . . . You can see it starting with some people at the age of 40, shall we say. They've got 25 years of life left, and they will continue to do a job that you can see that they could do so much more.'

This attitude can pervade a whole institution. Researchers tend to blame the 'enormous inertia' of some R&D organizations on bureaucratic factors, but they also appreciate the dangers of 'growing old together':

'People do get older, and they get terribly stale . . . . In other parts of [this organization] we see an awful lot of staleness. People seem to get old together . . . . They used to talk about the work, and then they got down to talking about the garden, and in the end they don't talk to each other at all.'

Indeed, research managers know that they may have to kill off projects arbitrarily before they become fossilized (Lemaine *et al.* 1972). This is one of the reasons why the Medical Research Council has always had a policy of closing down a research unit when its director retires and starting afresh with a new unit on another topic. It is a commonplace of contemporary science policy that university departments and research establishments need continual infusions of 'new blood' if they are to keep active and alive in a period of institutional stagnation or retrenchment (§2.5, §3.5).

The desirability of a regular supply of fresh faces is a general principle that is thought to apply to almost any social institution. Many people believe that R&D organizations, as such, are peculiarly prone to institutional sclerosis (§9.1), and cannot keep going satisfactorily without a relatively high proportion of young people. It is widely assumed that the majority of scientists tend to 'go stale', in mid-career, regardless of their institutional environment.

But this assumption may be fallacious. Individual aging is not necesssarily the primary cause of individual stagnation. It may be due simply to lack of personal or institutional *movement* (Bailyn 1981). As we have seen (§4.8), there is little real evidence that the *technical* capabilities of researchers decline with age. Scientists themselves are ambivalent on the subject. On the one hand they are scornful of:

' . . . this theory that civil servants have that scientists get burnt out . . . . I think this is simply a thing dreamed up by the civil service to avoid promoting scientists into administration'.

On the other hand, they qualify their belief that scientists don't 'get burnt out', with the proviso — 'if they were ever any good', thus indicating their opinion that those researchers who show this effect should not have been recruited in the first place.

What scientists do acknowledge about their colleagues — and, by implication, about themselves — is the possibility of a change of motivation with age:

' . . . it's fairly natural that, as people get older, their interests will change . . . . I don't see why, over your whole lifetime, you should only be interested in one narrow topic . . . . I don't necessarily think that [people who are said to have burnt themselves out] lack the ability to contribute to the science — merely that something else fascinates them more. Most of my motivation for what I do is what fascinates me'.

In other words, they attribute 'staleness' to a loss of *commitment* (§3.4), rather than to a decline of competence.

The basic assumption of academic science is that a researcher can only do good work if he or she isfree to follow up his or her personal scientific 'interests', whether these continue in a narrow specialty or whether they change markedly over the years. Indeed, the relatively small proportion of active scientists who have 'migrated' to other specialties (§1.6), often say that they did so out of 'boredom' (Gieryn 1979), as if no other factors such as career advancement affected the decision. Nevertheless, the evidence at many other points in the interviews suggests that the vocational commitment of a scientist is to the research process itself, rather than to the solution of problems in the particular specialty that they happen to find 'interesting' at the moment.

5.7 The attractions of a change

Change is seldom positively desired. Very occasionally, a scientist says:

'I am slightly different from the others here . . . in that I deliberately, through my career, tried to broaden myself . . . I [made a succession of changes from one field to another in a large discipline] deliberately. That's because I wanted to become a very broad individual.'

Another may say:

'I thought that I had been working on [−] rather long enough. It was not that I was bored with the work, because the work can be as interesting as you care to make it, but for a variety of reasons I decided that a change would not be untoward.'

A scientist may even feel:

'I have reached a stage where I would not mind doing almost anything else . . . . If I had the opportunity at this stage, I would quite happily drop what I am doing and go on completely even in an administrative sense.'

There is also:

'quite a distinct character who deliberately moves about every two or three years, and gets tremendously rapid promotion . . . . They have done it purely and simply by moving at very short intervals . . . . They tended not to be our better scientists. They start off [in their early 20s] as a deliberate policy. They see it as a way of getting to the top, usually in administration at a high level'.

Generally speaking, however, researchers suggest that they have been *pushed* into a change of subject by circumstances. They say that 'the project around me was crumbling, and one had to look for pastures new', or that current research had effectively 'reached its goal', or that 'it was a most soul-destroying job . . . once we had developed the techniques', or that their project would have 'had to wait for a long time before it got the money' for further technological development. And when a change is proposed, to 'turn something down . . . is really saying that what the management sees, perhaps quite correctly, as the only use of me in the company is no longer there'.

Nevertheless a radical change of subject can sometimes look very attractive:
'When you know things are going to change anyway, you think, "Well, let's make a clean break". This is a new challenge. OK, there are risks the whole thing may collapse in a heap . . . . So it was a gamble. You could see that it was interesting, and . . . it was a new challenge coming up . . . . It was put to me "Would you like to take this on? There is a new area. There is a growth area", and it was sold to me as a challenge. To do something . . . "You made your mark there; you can stop doing that now, start something new. Mid-career! Very positive break! Are you prepared to take on a new challenge?". And I did. I think I will, in the end, think I made the right decision.'

Seen in retrospect, a change may look even more attractive:
'It's been a challenge. It's been hard work. It's been fascinating. It's given me a bit of everything, and I have no regrets about it . . . .'

This challenge is not so much professional as vocational. It appeals to the personal commitment scientists feel that they must give to their work (§3.4). A scientist may not realize that he or she is 'getting into something of a rut', but . . . .
'Then, all of a sudden, somebody says, "Well, come to this meeting. We are going to talk about something different", and suddenly you begin to get more interested. So I would say . . . "Expose yourself to a few different situations, and see what happens".'

Merely 'going into a new environment, learning about a new [topic] ' is found to be 'stimulating' and ' exciting'. Getting involved in a novel project was seen as a 'very interesting exercise', even though it was not eventually successful:
'If there s a major objective which you find very exciting you don't mind being moved.'

Quite generally, scientists believe that:
'If you are of an enquiring frame of mind, with a fair degree of intellectual adventurousness, taking a fairly major step is quite exciting.'

But they know that it is often more difficult and usually less exciting to move into an established field (§5.3) than it is to take up a completely new theme, where 'there was no expertise here really', or there was still 'scope for changing the broad strategy or the details', or where 'we had to create our own knowledge base at the forefront all the time', or where 'it did not take you long to straggle to the front, because there was not a vast amount of literature in a lot of the areas'.

The attraction of a relatively unexplored field on the boundary of a traditional discipline (§1.2) was that even 'adequate and reasonably sound' research might reveal 'really interesting and important discoveries'. Thus:
'I was glad that I had entered into [−] because compared with my old area I could make progress fast. This is true of many new areas. You make

progress fast. That gives you more confidence. And it gives you more
enthusiasm, and it sets the ball rolling, so that it's difficult to stop.'
Most scientists would thus agree that:

'. . . if you move into a field that is just developing, then it is relatively easy
to make that change, because you can quickly become abreast of the
developments in that field. If you are making a switch to a field which has
been established for a long, long time, then it is far more difficult to make
the change. It is far more difficult to . . . learn the techniques and get abreast
with the literature, and the current methods . . . . If you move into a field
which is a new field, you can develop as the new field is developing itself'.

But there is also agreement that an experienced scientist coming into an established
field can be surprisingly successful, because he or she will 'tend to bring their own
expertise into the field, and because it is so new their own ideas are carried forward
a long way'. It is not necessarily a handicap not to be a real specialist in a subject
(§1.4):

'I had a bit of luck, I suppose. Not knowing anything really about the detailed
literature of the field, I guessed [the solution to the problem]. Of course I
guessed that from ignorance, because the people in the field knew that it
would not work − it was part of the folklore. So coming in from outside,
sometimes, there is a benefit of ignorance.'

In another case:

'We found the level of science was very poor indeed, and when we entered
it we could see mistakes that people were making, and were able to pick these
up straight away . . . . It takes an outsider, often, to see all the problems,
if you can just look at it afresh: whereas somebody who is in it can't see the
problems.'

Considerations such as these do not weigh heavily against the evident difficulties,
doubts, and disadvantages of a radical move. But they are very important in the
present study, because they take the sharp edge off these apprehensions and
indicate ways in which scientists can be encouraged and supported through the
transition period (§8.5).

5.8   Institutional and geographical mobility

Because their work is so specialized, science, engineering and technology workers
are less occupationally mobile, but more geographically mobile than other people
(Gleave 1985). Any substantial change of research field or organizational role is
almost bound to involve a change of working site − whether into an adjacent
building in the same establishment, or into another establishment, in another R&D
organization, 500 miles away. Such a move may raise daunting practical

problems. Career change in science is crucially influenced by all the factors affecting *institutional and geographical mobility* in society at large. Thus, for example, American scientists are said to be willing to move around the country much more, from university to university or from firm to firm, than their British counterparts: the main reasons for this are probably much the same as those that make skilled and unskilled workers in all callings much more mobile in the United States than in Britain (§9.2).

These 'extra-professional' considerations are so individual and multifarious (§4.1, §5.1) that they could be dealt with only very superficially in the group discussions. It is true that 'you can never satisfy personal problems; there are always personal reasons for being in one place rather than another at certain times in your life'. Nevertheless, they can only be treated here as boundary conditions within which individuals must make their diverse decisions.

Some of these conditions are obvious enough. Many people, for example, would resist a move from the South of England to the North of Scotland, for reasons of climate and of access to the amenities of metropolitan life. Such resistance might be compounded by the cost and inconvenience of visiting parents, or keeping in touch with friends, at such a distance. On the other hand, it may not be easy to move to a more favoured region of the country because of the higher cost of housing (Morris 1983). One of the major restraints on individual geographical mobility nowadays is the difficulty of finding a satisfactory job for an employed spouse. Most parents are anxious about changes in the schooling of their children, especially in their teens when critical examinations are coming up – and so on. There has been a big change since the time 'years ago', when 'the man would have gone where the work was, and the work he enjoyed doing, and the family would have to drag along with him'.

It is noticeable that research establishments are often very pleasantly sited, in quiet country places with reasonably good access to intercity transport, not far, perhaps, from a major university. People who have 'put down their roots' in such 'comfortable' spots for a number of years are naturally very reluctant to move elsewhere, especially if they have become deeply involved in the life of the local community. It is sometimes said, of course, that:

'you get people who take up interests outside their work, and maybe get too involved in that: that could lead to staleness in their scientific work',

but Bailyn (1977) found that people who 'accommodated' to work in order to become more involved with their families were not noticeably less effective or valuable to their employers.

A scientist who is quite ready to make a radical change of research field may jib at combining this with a major domestic upheaval. This objection does not arise if the change does not involve moving house. For this reason, researchers working in a large establishment with a diverse programme are potentially much more intellectually mobile than they would be in a smaller institution with a more limited range of activities (§7.3). Moreover, as Marshall (1969) points out:

' . . . it is not difficult to change emphasis or orientation in a multi-discipline and multi-purpose laboratory like Harwell . . . but it is more difficult to bring about change in a single discipline, single objective laboratory: such laboratories have many attractions but it is difficult to know what to do when their objective or mission has been achieved'.

But for the individual involved, even transfer to another 'division' on the same site is experienced as a change to an unfamiliar social venue. Thus, the alternative of moving to another establishment within commuting distance in the same geographical area may be just as acceptable.

At first sight, this alternative often looks quite feasible, since there is a tendency for R&D establishments to cluster in certain regions of the country, such as the Thames Valley or the neighbourhood of Cambridge. But any change of employment raises many practical problems. A move from one sector of R&D activity to another (§2.4) − e.g. from a government research establishment to private industry, or from a research council laboratory to a university − may involve a serious sacrifice of pension rights, not to mention possible loss of seniority and entitlement to permanency of employment. Indeed, these administrative barriers to movement between R&D organizations are so notorious that they are often cited as a primary cause for the characteristic 'immobilism' of British scientists.

But this is not the whole story. Even within a single R&D organization, such as one of the research councils, it may not be normal for scientists to transfer from one establishment to another, especially if this would also involve moving house (§9.1). Such transfers can, of course, be arranged, in particular cases, without loss of pension rights, etc. but only if a post becomes vacant at the right moment, etc. etc. In the past there were often no regular administrative procedures for facilitating such moves: nowadays, with staff numbers contracting, the openings for them are very few.

This brief and schematic discussion of 'extra-professional' factors fails altogether to do justice to the importance of such factors in shaping the lives of scientists and other technical workers. As Hutt (1981) remarks, we have very limited hard information about the 'occupational, functional, locational, and employer-related' dimensions of career change, all of which are interdependent. But we shall return to a number of these points in Chapter 7, when we come to consider the direct effect of managerial policies and administrative practices on specialization and change in scientific careers.

6

Life beyond research

6.1 Career paths out of science

Academic science scarcely admits of life beyond research. The reputational career of a scientist (§5.3) is summed up in a *curriculum vitae* lived out entirely in the laboratory or the study, the lecture room or the conference chamber. But in the normal course of their *organizational* careers (§3.3), many scientists are invited, or instructed, or themselves seek, to move into jobs that do not involve them personally in research. In such organizations, it may even be asserted that:

‘ . . . the scientific career where one saw oneself as going through as a scientist the whole of one's life was a nonsense . . . the services have got it right . . . they pension people off at 40 or 50, about when their useful and active life has come to an end, and . . . that is what we should do in science . . . . We should have a system of early retirement . . . . Scientists should go and do something for which their career has fitted them, but not expect that they are going to stay in the Lab and do research until the day they retire’.

Such moves must obviously be taken into account as possible alternatives to a change of specialty within research.

In most R&D organizations nowadays, a scientist who has reached a certain grade, e.g. in the Scientific Civil Service (cf. §.3.3), the rank of Principal Scientific Officer, is bound to consider the possibility of moving out of technical work into research *management*. Scientists in mid-career often talk about their career prospects primarily in such terms.

But genuine managerial posts within R&D are relatively few in number, and almost always high in rank. Indeed, they tend to monopolize the senior positions in the organization, so that, except for a few outstanding research scientists and design engineers, ‘going into management’ is normally an unavoidable step on the ladder leading to higher levels of pay and esteem. Most of the way that scientists talk about such a possibility is really about their hopes and fears for personal promotion (§7.5). Whether or not such a prospect is realistic for a given individual, it naturally plays a very important part in the attitudes of scientists towards this kind of work.

For the majority of scientists and engineers, however, mid-career promotion into a managerial position within R&D is not an eventuality on which they can firmly count. If they are contemplating a move out of active research, they need to think about a variety of other jobs for which their research experience might fit them. They would naturally consider the possibility of undertaking *administrative* (§6.3), *technical support* work (§6.4), or *teaching* (§6.5) within the R&D system, or of taking a technical, administrative, or managerial job in the wider world of industry, commerce or government. On occasion, any one of these may look attractive as a realistic alternative to a change of specialty within research. In the present chapter we look briefly at the way in which working scientists see them as possible career paths.

6.2   Research management

The traditional view of science as a *vocation* (Weber 1918) has no place for the R&D *manager*. Nevertheless, nowadays:

'when you are talking about careers, then to most of us that must mean, ultimately, the possibility of going into management'.

This does not apply, of course, in academia (§2.4). The senior staff in a university, that is, the professors, readers and senior lecturers, are not considered to be professional "managers" in the usual bureaucratic sense. They may in fact be expected to spend a great deal of their time sitting on committees, drafting papers and dealing with problems of people, equipment and finance (§6.5), but this is only one part of their "job", and the extent of such responsibilities is an inappropriate measure of their career success. Quite the contrary: what they are supposed to do, if they possibly can, is to go on with their research. Like their more junior academic colleagues, they are selected primarily on their reputation as scientific specialists, on the assumption that they will treat such an appointment as an opportunity for enhancing their standing as such.

This is the model for the 'Individual Merit Promotion' scheme (§3.3, §7.6) operated by the research councils and the scientific civil service. In these organizations, the primary criterion for promotion up to the PSO level is usually direct research performance, often measured in terms of publications, just as in academia (§5.3). Outstanding performance according to the same criterion can then earn promotion to higher rank without specific managerial responsibilities. Many large industrial R&D organizations offer a similar 'dual career ladder' up to the rank of 'senior scientist' or 'senior fellow'. But the proportion of such posts is very much smaller than the proportion of senior posts in academia itself. Various long-term and short-term forces (§2.1, §2.5) are pushing all the R&D organizations in the research council and government sectors away from the 'quasi-academic' model towards the more 'managerial' industrial pattern, where:

'it is generally accepted that a person's lifetime at the bench, compared with university, is considerably shorter, and that there are continually other options coming up, and that in fact there is often quite a large turn-over . . . of scientists coming through, others moving on'.

Nevertheless, very few science graduates seem to enter research with the initial intention of becoming senior managers or administrators (§7.2). But 'in their 30s', before they have reached 'mid-career', many begin to ask themselves whether they should take positive steps along the 'management route' in their organization:

' . . . there is actually a key stage, after four or five years [from recruitment as an a researcher with postdoctoral experience] . . . . You are doing the job as a research scientist. You are beginning to think you are aspiring to a senior scientist position, and not really interested in knowing people, or you are beginning to drift away from the detailed science, and wanting to be a project manager, or being given opportunities by the management towards that position. That is where the two sides are looking at each other — after that sort of period of time, it gives one a turning point'.

Even if they are doing good specialized research, some scientists already 'don't visualize being able to maintain that sort of excellence until the age of 60', so

'they test themselves on their management abilities, and then if they take to it, or if they are adept at it, then they consciously choose to further their career on the management side, forsaking their own individual contribution to the research, and take their job rewards in the management'.

Some researchers thus acquire this ambition or are secretly identified by their bosses as potential 'high flyers', quite early in their careers. From then on 'running big teams and running around, that's more important than doing research', and 'you've got to move around, you've got to move establishments, you've got to become very broad-based'. Experience of a diversity of organizational roles counts for more than specialized technical performance (§5.4).

But most scientists are reluctant to 'give up their research interests and cap-ability' at this stage. As Sofer (1970) puts it, 'like most professional people, their predominant wish for the future is to stay in their present function, with promotion to a higher rank'. Some, indeed, deliberately reject managerial responsibilities, knowing that

'I have reconciled myself to a certain level of career development because I prefer and choose to maintain a personal group working actively in research, rather than pursue management actively, and I regard that as a restriction on my own career.'

But many of them find that they are carried towards the 'management route' by the natural evolution of their careers.

This movement may be signalled by being asked to carry out a specific task:
'. . . . the section head decided to change the organization of the section . . . . He said "I would like you to come away from your pure research, if you like, and get in there and help sort it out". . . . But that meant . . . at a very early stage, me realising that I wouldn't be working at the bench all the time, and I would be moving a little way away from it, and getting into a managerial aspect . . . . I still maintained my own activity and interest in [my research specialty] through assistants, but not working at the bench'.

At some stage, a more 'extensive' (§2.3) research project may have to be undertaken:
'The most significant shift was from doing the fundamental research work, with a sort of a man and a boy for help, to running a larger project of ten or 15 people responsible to me, and I suddenly realized what I wanted to do was management, rather than doing the job myself, and I got a lot of satisfaction in having people understand what I wanted and then going away and doing it.'

But the same trend occurs automatically, through the normal course of promotion. The *reputational* authority of the research *leader* is inevitably transformed into the *organizational* role of the *manager*:
'[At the level of] SSO [management skill] starts to come in; by a PSO you are a manager, with a staff and project. By that level you have to be good at what you do [as a researcher] and you have also got to be a very good manager with staff, a very good public manager.'

The increasing 'collectivization' of research (§2.1) induces this transition at an earlier stage. The older staff in one establishment could remember a time when an SPSO was permitted to spend 50per cent of his time doing 'scientific work', or at least be the 'technical leader' of one of the groups under his control. Now, it is said, the research is fragmented into so many small commissioned projects (§2.4) that even at the PSO level 'they are being pure administrators and not at all scientists'. This 'bureaucratization' of research is accentuated by the bunching of staff in the middle ranks of the hierarchy (§3.5). In another establishment it was said that:
'whenever an officer is promoted to PSO, he is given a small section of his own. As the number of PSOs continues to grow . . . the science within the [establishment] is broken down into more and smaller slivers',
even though:
'some PSOs would be quite happy to get on with their research without having a large retinue of junior staff to look after'.

At whatever level it occurs, the final transition 'from bench work to management' is seen by almost everybody as a 'major turning point' in a scientist's career. It is:

'really quite a change to have to stop doing experimental work, and to rely upon somebody else to do the experiments, to tell them what to do and to judge who you could rely on to come back with the right answers'.

Whether or not, as Reif & Strauss (1965) argue, this transition is actually easier than taking up new research, it has strong affective implications. As they put it:

'For the scientist who has internalized the norms of discovery, and especially if he has earlier proved himself an outstanding discoverer, the transition between active research work and the shift into non-research activities is . . . fraught with personal difficulties and great demands on adaptability. The transition marks a critical turning point in the scientist's life and career: his scientific 'menopause'.'

Those who are directly faced with it find it:

' . . . a little bit abhorrent . . . . It's a sort of Peter Pan thing . . . . It's like thinking about dying. At some stage you have to become a manager. Horror of horrors! You are no longer the person you were . . . '.

On the other hand, they realize the danger that:

'sometimes, by wanting to stay where they are, they're automatically making sure that somebody who they used to look after is going to be looking after them − that somebody's going to jump over them − and that's difficult unless you can prepare for it . . . . It's not that their ability isn't sufficient to take them to the top: it's that their motivation really takes them along certain lines which almost denies them promotion'.

It is fully appreciated, of course, that not all scientists should make this transition. The scientist who 'could usually be found in his room with his feet up on a desk, reading a detective story' is not an obvious candidate for administrative office. For others, it seems 'a tragedy' to take them out of research:

' . . . because they seemed to have great ability to go on. Others quite clearly used their administrative position . . . in order not to do any work, because they haven't got any more work in them'.

Some people might be said to have 'alternative career options' − 'business or administration, versus academic', say − whilst others do not, although 'there will be some people who have an option and they don't know it'.

New skills are obviously required. In the present highly competitive climate of R&D (§2.5), a research manager has to be:

'politically skilful about selling your ideas in an accepable form, so that people can give you money . . . . If you have the wit or particular aptitude to phrase things in a way which is fashionable, you can do much more which is satisfying'.

Stress is laid on having the right social personality − on being able to 'get on with people':

'You can have a damn good scientist who is no good at the managerial side . . . . We have got [a person] . . . who is exceptionally good at science, but difficult with people.'

On the other hand, it is often said that the people who are paid to do management full time 'are very flexible . . . it doesn't matter to them whether they are managing a carpet factory, or whatever . . . they are dealing with people all the time'.

But there is another side to R&D management which is more relevant to our present study. The expertise required to run a team of researchers, however skilled and specialized they are individually, still has 'a very high scientific content', spread over a very wide field:

'My experience is that one is called upon to become more versatile as one gets more senior. At the start, when I joined the [−] unit, I had a series of small projects that I was following which were relatively narrow, and I could get into that field relatively easily, and read up and develop that, and publish on that particular aspect. And then one becomes a manager of small-scale projects, where one has to develop one's expertise a little more. Then one gets up to the situation where I am head of the unit. I get reports thrown at me which I am supposed to make critical comments on, on all sorts of topics [covering several scientific subdisciplines]. Then one gets into a situation where, if one became an assistant director [of the establishment], one is expected to make intelligent comments on reports in a much wider sphere [covering the whole discipline and its applications].'

' . . . There is a danger that when ADs have these reports placed in front of them, they feel that they should have to write something about it, rather than actually being capable of writing something about it. It might be a field on which they actually know nothing about . . . .'

' . . . The way . . . all scientific subjects develop, it is not always very easy to be in a position to have a report thrown at you from all sides on a lot of different topics, and to be in a position to be able to assess these, even at a relatively low level, because you are responsible for that work. Whether you like it or not, responsibility falls back on the person who is in charge.'

The essential prerequisite for middle management in an R&D organization is experience of research 'at the bench'. This is the only way to learn 'how to do research' (§4.7), and to 'know enough about what is going on so that you can interact' with the scientists whose work one is directing. But the typical research specialist moving into a managerial position is often quite unprepared for the job. He or she may have had only 'odd courses that are called management training, but really they are not − not in the professional sense of management, at all'. Not only do they come into the position as 'amateur managers', expected to 'pick up skills on the way': they also come with a very narrow domain of technical expertise (§1.3), which has to be broadened very rapidly to deal with the diverse scientific questions that arise in a research project that may span several disciplines (§2.3).

Scientists who have successfully gone through this transition do not underestimate these difficulties, but do not regret that they faced up to the challenge:

> 'I suppose over the first year I felt very strongly that I had reached my level of competence, and I was struggling in the sense that you get into that phase whereby you have got an enormous amount of new stuff to learn. Different pressures are put on you in terms of management, whereas you are fairly competent in science, and you think ''now this is where you are''. I guess with the experience of time you suddenly find that you can handle this, and you want to look at bigger and better things.'

There is a real satisfaction in being 'able to be involved in quite a lot of projects, and have some influence in some of these areas', even though this makes 'a big jump from the laboratory, away from any data base whatsoever'. There is the 'creativity that comes from pushing the people who are doing the work in a particular direction'. The theme of this whole book can thus be seen from quite another point of view:

> 'I see the disadvantages of not being the expert in the team. But I also see the advantages of being slightly outside it − all right, above it − because then you are looking down, and can actually ask the questions. That's important in good management. The guy has all this opportunity, all these experts running round beneath him. And if you can ask the right questions, then you can really make it tick along. So I don't feel any sort of . . . fears about experts − people who know more about things than I do. In fact, in many ways I look forward forward to being able to ask somebody a question who will give me an answer in two seconds.'

6.3 Administration

Throughout the R&D system, scientists try to avoid being 'lumbered with administration'. They say that their research is frustrated by the 'pressures to do other things, to sit on committees, to get on advisory panels . . . and other things that are not so rewarding and not so satisfactory'. One full-time researcher in a research council unit suggested that he would 'think very hard about' promotion to a professorial chair because 'I'd realize that it would more or less mean the collapse of research for me in general'. Indeed, the general assumption in academia (§6.4) is that if someone is not doing much research, then there is 'a responsibility to give him useful things to do for the community for which he is paid, and therefore that man probably gets . . . more administration' – and thus takes some of that burden off the shoulders of his more 'productive' colleagues.

But the scientist whose ideal life is '90per cent work [i.e. research], 10per cent administration', does admit that he would not be unhappy with 'an administrative rank where I felt I could have a real influence on the direction . . . it was taking'. In other words, scientists and engineers make a distinction between the *line management* of R&D, with all its responsibilities and prestige, and other less visible forms of *administration*.

In fact, most R&D organizations have many other administrative jobs besides the obvious posts in the official managerial hierarchy. The pressures on the system (§2.5) have generated a need for much more systematic 'programming and planning' of R&D, for 'forward crystal ball gazing', and for 'commercial officers' to 'market' the research products and capabilities of each establishment. Scientists in mid-career, with good technical qualifications and considerable experience of research, are often very suitable candidates for such jobs. They thus become 'scientific generalists', and 'go into matters of strategic assessment, project assessment, and project management', in some cases carrying out work that is 'more analogous to the sort of work that technical grades carry out in the Civil Service'.

In industrial R&D, moves of this kind seem to occur quite naturally:

> ''I decided I wanted to change . . . because I like to feel I am continually broadening my horizons, and an opportunity came up in the personnel department, which was in manpower planning, which I had become interested in through having served two years as a representative of our middle management on the manpower representative organization. So I moved into personnel, and did work really setting up manpower planning techniques of use to management.'

But in some research council and government establishments, such changes of professional role are not, apparently, quite so easy. For example, a biologist who had become interested in 'human development . . . psychology, philosophy . . . what's it all about, and what happens to people' thought it unlikely that he would be able to go into 'something like training, or even personnel work, or even things like . . . psychiatric work'. Although administrative work often occupies almost all the time of a number of fully-qualified scientists, it is said to be 'very difficult to identify a continuing activity apart from research', and that 'there is effectively no non-research career structure'. The senior management 'feel that you should be doing it, that it's a good thing, but they begrudge the time spent on it':

> 'The difficulty is the unrecognition for doing non-research activity . . . .
> If you get true recognition for it, then it's probably fine, but I am quite sure that there would be people around who are saying "You are not publishing

x papers a year. What is wrong with him?'', and I think this factor still prevails to a certain extent.'

The surprising thing is that this attitude is to be found in the Scientific Civil Service, which is a segment of the largest administrative body in the country. How is it that administrative work is so undervalued and that scientific civil servants do not move freely from employment as technical specialists into the wider reaches of the public service? It is often remarked that few of the top university graduates in science and technology compete for entry into the Administration Group of the Civil Service, by comparison with graduates in the social sciences and humanities. The Administrative Civil Service thus lacks people trained to understand the scientific and technological aspects of modern government. Surely this deficiency could be met from amongst the many capable and experienced scientists already employed by the central government?

This is the rationale of the Senior Professional Administrative Training Scheme (SPATS), which is 'designed to allow able specialists who have obtained early promotion to Principal level to gain experience in policy making and management' (Holdgate 1980). A PSO accepted into this scheme is appointed for a short period to an administrative post outside the normal scope of the Scientific Civil Service, with the possibility of moving on and up within the Administration Group. The policy of temporarily posting a promising scientific officer away from a research establishment to departmental headquarters has a similar objective. This is necessary, because most of the regular work of the administrative civil service is not done in the research establishments at all. But such a move may be strongly resisted. One researcher reported how, at one stage in his career:

'there was a big push to move people to headquarters, and I then adopted two postures. One was one of bland incompetence, and the other was that of a low profile. The incompetence was so that I would not get moved to headquarters, and that involved wearing dirty pullovers and no ties for a two-year period . . . . The low profile did not work at all, I am afraid, because I was the first one [the director] moved, which in fact was . . . one of the luckiest breaks in my life'.

But even when the move is voluntary, it does not necessarily lead to a permanent change of career:

'So I was looking around, and thought I would [go] off to headquarters to do an administrative job for two years. So I took that [opportunity] to have a look round to see what the administrators do. And after two years, I would then have to decide whether to stay or come back here, and eventually decided to come back here as a scientist, rather than the various administrative jobs possible here as well. So I decided to go back to the bench, essentially.'

Although a career in administration often 'looks rosy', many scientists say:
'I enjoy science, I would rather do what I enjoy, and stay several grades
behind in my career'.
There is thus a widespread view that 'a move to an HQ posting or to a SPATS
course places the individual scientist at a disadvantage compared with one who
continues to display scientific abilities in a narrower sphere' (Holdgate
1980).

There can certainly be little enthusiasm for a diversion from a promising
reputational career if, in fact, one is only 'given this or that report to write, or what
one might call irrelevant busy work, and sent away in two years'. The SPATS
procedures were intended to be more systematic and challenging, but these were
also strongly criticized by one scientist who actually went to interview and was
accepted:
'I turned it down because it was asking me to commit professional suicide.
The way in which the scheme worked, if they accept you, is that you are
thrust into a job which is intended to be demanding. Now the scheme is
designed, in fact, to have you thrown into an administratively demanding
role as a scientist with a new parent organization . . . . That I could understand, so I said "OK, so I am a bit older than most . . . what have you got
to offer?". And the two things they offered − one was the job I left in
[−] . . . I did not really want to go back and do that, and the only other thing
they could offer was the other side of contract work − dear me! − neither
of which I thought were going to help my career one jot.
So I said "How do you go about finding [jobs] for candidates?". "Well",
they said, "it's quite easy. We sort of make a little CV out, and we will send
it around all the departments and see who wants [the candidate]". Well now,
I know darned well from here that if we had a letter from headquarters saying
'we have got a guy; you can have him for 18 months' [we would] look around
for the job that nobody else is going to want to do, and that really nobody
wants done, and put him into it . . . . Now had they come back and said
"Great! We have a function for somebody with your sort of scientific
background for 18 months, and it's really going to work the pants off you",
right, I would have taken it. But this idea of casting myself loose and drifting
round: there is no way that at 40 or 42 that I was prepared to take the risk.'
This weakness in the SPATS procedures is sometimes attributed to the fact
that:
'there has always been a reaction against scientists in high places in the Civil
Service. I have seen it at first hand, and so I think there is a lack of enthusiasm
to really making a difference between a pop scheme on paper, and make it
actually go, which means they have to give encouragement and confidence

when they get there, and are not made to feel that they are really a sort of second class citizen, working with the elite, which is the impression which has been given in the past'.

The scientists strongly resent the elite status of the Administrative Group:

'We always feel undervalued as against administrators, who really could not exist if you did not have the science to administrate in the first place . . . for example, somebody pointed out to me the other day that [in administration] a Principal at 27 or 28 may not be uncommon, which is incredible . . . . We are at least of the order of ten years behind that, and that is irrespective of how hard the individual works, or his capability.'

and console themselves with pride in their own profession:

'It's not like the rest of the civil service, which is really a bit of an Aunt Sally, trying to do a job for the government, yet the general population doesn't really feel that it wants that job done. At least we do have a well-defined structure, and a reason for existence, and we can see that we've achieved a good job, which I think in other areas of the civil service isn't necessarily so.'

But the failure of the SPATS to produce 'scientific generalists' for the administrative civil service is not due simply to professional tribalism, or to the perverse ambitions of scientists to remain narrow specialists. As we saw in §6.2, many scientists who have, in fact, been quite successful as research specialists are equally successful when they move into much broader managerial roles within the R&D system. Indeed, as Holdgate makes clear, the avowed aim of SPATS is to 'help to equip [able specialists] for the senior posts that some will later come to occupy', and '*not* to turn specialists into administrators'.

The real point is that the job of the typical administrative civil servant is not that of a 'manager' in the usual sense, for it does not necessarily involve directing the work of organized groups of other people. It may be 'generalist' in not requiring close acquaintance with any single professional technique such as a scientific discipline or branch of engineering, but it is highly specialized in its own way. This is because it deals with the domain of 'policy', rather than with the more immediate and tangible domains of material objects, specific programmes of work and the careers of particular individuals. It thus requires informal skills that are only picked up after considerable direct experience of the procedures involved. A graduate entering the Administration Group has to learn this expertise through a period of apprenticeship in a number of posts, often in different government departments. It may be even more difficult for a scientist to move sideways into the administrative civil service in mid-career than to take up research in a new scientific discipline, where at least the 'method' of research itself (§4.7) would be familiar.

6.4   Employing technical skills in other ways

Research scientists have many technical skills (§4.5) that can be put to good use in other ways. For example, almost all experimental R&D makes extensive use of sensitive electronic instruments and large-scale computing facilities. The provision and maintenance of these instruments and facilities is largely in the hands of technical staff, trained as electronic engineers, program analysts, etc. Nevertheless, it is not uncommon for a research scientist to become heavily involved in one of the *service* functions of a research establishment, and eventually to make this their career.

Such employment may not rank highly in the hierarchy of esteem within the scientific world, but it has its gratifications. Many research workers were drawn into science originally through their interest in the technical practices of research, and they get real satisfaction from 'tinkering' with apparatus, or designing novel instruments or programs. In an organization with large instrumental needs, this can lead eventually to a position with considerable managerial responsibility:

> 'Eighteen months ago I went back to the instrumentation side . . . where I suppose I am most at home, although now of course I am not involved on the design or the research side so much as controlling groups and teams of people who are engaged on that sort of thing . . . . I could not now see myself getting down to the more basic research and investigation and design work that I could have done quite readily say ten years ago . . . . I [act] as a fairly effective filter in a system, which controls the excessive zeal of some people and also perhaps pushes along some of the other groups. There are a very large number of projects that I am looking after . . . . I have a team of about 20 people now working for me, all with different expertise.'

A researcher in mid-career may well find that a move into this kind of employment is preferable to a move into another field of research, or into a non-technical administrative role.

Some R&D organizations exist primarily to provide technical services to industry, to agriculture, to local authorities, or to the general public. A scientist whose research trail has already come close to this interface may thus be very well qualified to pass through it and carry out this 'external' service function. Yet this does not seem to happen as frequently as might be expected. The corps of researchers is commonly separated professionally from the corps of technical advisers, so that there are no regular arrangements by which a researcher in mid-career can make this transition.

Nevertheless, in one major organization where all technically qualified staff belong to a single corps, it is quite normal for people to move from research into an internal or external service function, and back again, in the course of a career. In many organizations, this is done specifically to give people picked for eventual senior managerial positions the experience of working in a 'staff function', where they find they have 'a lot of repsonsibility, but almost no authority' (Wolff 1980). But it may also be done systematically, as managerial policy (§7.3), in order to avoid the problems of undue persistence and career rigidity with which this whole book is concerned. Such moves may, of course, involve large jumps across the disciplinary map (§1.1):

'On promotion to PSO, I was sent to [a place far from London], and there are two aspects of this which were a complete change. The first one, of course, is being a long way away from Headquarters, and so having no direct supervision from above, and being my own man . . . . The second one was transferring first of all from the research to the services side of the [organization], and secondly transferring from a mathematically-oriented world into one that was governed really by physics . . . . I immediately became very much aware that I had been a mathematician, and so it took me − I would say two years − to come anywhere up to scratch as being useful . . . . [It involved] a complete change of literature as well . . . . [I had to] dig out different journals and catch up on five or ten years of work that someone else had done . . . . There was a biological input as well: I had to take an interest in [a branch of physiology], for instance . . . and so there was a whole new realm here which taxed me enormously − so much so that after about six months I got really fed up. But I stuck with it, and in the end [it was very satisfying]'.

What is more instructive, however, in this particular case is the way in which a research specialist had to adapt to a more general organizational role:

'A lot of it, of course, was also relationships with people outside the [organization]. I think in the research atmosphere that I had been used to I didn't really see anyone outside. With the job I then had, I had to do a lot of public relations work, get out and see quite a lot of people from a very broad spectrum. So it really was a complete change.'

Q.[What made it so satisfying?]

'It was a mixture. I do now quite enjoy doing the relationship type of work, and I still carry on that, in fact, in my current job. But I think it was the basic change from seeing myself purely as . . . really only . . . a mathematician, into someone who could see a lot further into the other kind of work that was going on in the [organization], getting a grip of the rather broader issues.'

The same message comes from scientists in other public sector R&D organizations whose jobs have moved towards external service functions:

'A. 'I found it quite interesting, actually, getting involved with these people, and getting the interaction from outside. So I don't think this kind of work is always . . . — it brings certain advantages as well.'

B. 'Then you get a broader insight, interaction — see other problems from other people's point of view. I would change the work I am doing, in future. I think I would like to move further into that side of work in the [organization] — but not for ever.'

The trend towards commissioned research, and the privatization of a number of R&D establishments in the research council sector (§2.5) has pushed many researchers into semi-technical jobs, such as seeking research contracts or looking for sponsors for the development of technological innovations:

'In 1970 . . . what was called the [Establishment] Marketing Forum [was set up]. Marketing was a dirty word as far as I was concerned. I knew nothing about it, but it seemed to me that this was important and interesting. The marketing of the results of research seemed to be a stimulating thing, and so I just thought, "what the hell, I'll go and join this unit and see what it's like" . . . I went and told my Division Head, who was shocked, but I made the break, and dropped all research, more or less, within a few weeks . . . .'

Q. It's like emigrating to Australia, isn't it?

'A bit like that, yes . . . and you're not very highly regarded, you know — a bit of a turncoat.'

Q. Did you have a return ticket?

'Well, no, I did not have a return ticket, but I think that this organization is very good in this respect . . . . [It] can give people the opportunity to make such changes, and I reckoned that if it did not work out, if after a year I was sick of this, and was not any good at it, I could . . . go along and they would sort something out for me — probably in computing.'

In effect, many research scientists in the public sector are beginning to follow a career pattern that is well established in the private sector (§2.4). A researcher in an R&D establishment of a large industrial firm does not expect to stay 'at the bench' throughout his or her career. Quite apart from the hope of promotion to a managerial post (§6.2), there is always the possibility of moving sideways into a technical job in another part of the firm.

This may happen in a variety of ways. For example, 'some people, in some divisions, when they come up with the ideas, . . . follow them through, through the different areas' — that is, as was said in another firm, 'from the development stage right through to the market place, including the setting up of quality control procedures which we then handed over to production'. A scientist involved in this process may stay on in marketing, or get interested in quality control, or even end up running a manufacturing unit. Research scientists are often seconded to production units to deal with unusual technical problems, and settle in as resident trouble-shooters. At a more senior level the firm may need a co-ordinator to deal with a general technical problem, such as environmental pollution:

'I am learning a completely new role . . . . I started by sitting with Nellie – the man who had been doing the job. He is retiring now. I had an overlap period with him which has been exceedingly useful. And then it's a matter of going around the company, to all the company plants, to every factory, to talk to the works manager about what their problems are in environmental pollution, to talk to the people in the laboratory as to what contribution they are making . . . and hoping eventually to pull this lot together, to co-ordinate it . . . and at the same time to start making new contacts outside the company with the government bodies – the people who are passing the legislation on pollution – to make sure that we can influence the legislation in the sense that it turns out to be something practical that industry can meet . . . . So it's a matter of making contact with these people outside the company . . . . It's a job which involves science and technology and legislation and poitics, and somehow you are having to bridge these, and interpret one for the other.'

As the above case shows, the technical components of many such jobs in industry are highly diversified (§4.4), and are eventually submerged by more general administrative considerations. An industrial R&D establishment is always associated with a much larger manufacturing and commercial organization, to which it is closely linked by common tasks, staff with similar qualifications, and common managerial control. A career trail that began with entry as a PhD doing highly specialized scientific research may thus 'drift' across the organizational chart of the firm, without any abrupt break (cf. §1.7), until it finishes in an office at Headquarters dealing with personnel, or sales, or advertising, or public relations. The 'research laboratory' thus serves as a source of technically trained personnel, often of very great ability, who move out after a few years into the other operations of the company.

By contrast, the typical government or research council establishment is a relatively isolated institution, which is not closely associated with any larger non-scientific organization. It thus has very few administrative or technical openings for scientists moving out of research. This is why, on the whole, the problem of undue scientific specialization seems to be so much more serious in the public sector than it is in private industry and commerce (§9.1).

6.5 Teaching

Scientific research has always been closely connected with the teaching of science. The traditional career pattern in academia (§2.4) is to combine these two functions. This is achieved by employing people officially to teach, but appointing and promoting them on the basis of their research. The sole contractual obligation of a scientist on the permanent academic staff of a university or polytechnic is to teach a specified discipline or subdiscipline (§1.1): nevertheless, he or she is also expected to work very hard indeed at research, and time is allowed and facilities are provided to make this possible.

Nowadays, it is true, this traditional combination of functions is beginning to come apart (§9.1). On the one hand, many people are employed in the academic sector as full-time researchers. Some of these are relatively young postdoctoral research assistants, but quite a number of fully-qualified research scientists are now being forced by a shortage of academic jobs to continue into their 40s through a succession of temporary appointments which do not include any teaching, even though they are located in universities (§3.3). On the other hand, many scientists on the staffs of polytechnics have such heavy teaching and administrative commitments, and such limited research facilities, that they have very little opportunity to undertake the research for which they are also qualified.

This separation of the professional roles which many scientists had reckoned to combine is highly relevant to our theme. It has not come about by a naturally evolving division of labour within academia, but results directly from the abrupt economic and administrative stresses laid upon the British system of higher education in the past five or ten years (§3.5). It indicates administrative rigidity rather than institutional efficiency. Many of the people involved are well qualified, highly competent and eager to do both teaching and research, and the work is there to be done. But the idea that some of these tasks could be exchanged and shared out amongst them is taboo: the 'binary' line between the universities and the 'public sector' institutions of higher education is still too sacred to be crossed in that practical way.

This has to be said with some force. The 'collectivization' of science (§2.1) has not yet made obsolete the long-established practice of combining undergraduate teaching with advanced research. Difficulties can certainly arise with very 'urgent' or 'extensive' research projects (§2.3), which have to be undertaken by R&D organizations employing full-time researchers. There are the complications of funding very expensive and diverse research facilities. The practical exploitation of the results of 'pure' research can sometimes lead to serious conflicts of interest concerning patent rights, etc. It is not always easy nowadays, within a small department, to meet 'one's responsibilities to teaching and to cover the field adequately, and at the same time to optimize one's scholarship by focussing'. These are only some of the problems of maintaining research activity in an institution of higher education.

But there is no reason to suppose that separating academic scientists into career teachers and career researchers will make these particular problems easier to overcome, or make universities more adaptable to changing educational, scientific, technological or commercial needs. On the contrary, the practical capability that academics have, as individuals, of changing the balance between their research and teaching functions adds a valuable dimension of flexibility to all academic institutions − a flexibility that is not shared by organizations dedicated solely to research.

It also adds a valuable dimension of flexibility to an academic career. Academic scientists are trained, are expected, and themselves prefer, to be active in research; but if perchance the research runs out they already have all the alternative employment they could want in the lecture room and teaching laboratory. Many university scientists do indeed make this move in mid-career. In some cases this is due to circumstances beyond their control, such as a major cut in the funding of their research specialty. In other cases, they simply fall behind in the intense competition for external funding for their projects (§2.5). In other cases, again, they fall into the trap of 'undue persistence' in their specialty (§5.6), and lose confidence in their scientific ability or lose interest in research itself. Whatever their reasons for having dropped out of research, a considerable proportion of the older acdemic staff in university science departments spend most of their time teaching undergraduates, or carrying out the many quasi-administrative jobs associated with higher education, such as running practical laboratories, setting examination papers, selecting students for admission, or − alas − sitting on committees to manage all these functions.

It is easy for the managerial efficiency expert to scorn some of this as make-busy activity. Indeed, academics continually complain of the burden of unnecessary administrative work that they put upon each other, or that is put upon them by the 'bureaucrats' in the Senate House, the UGC or the research councils (§6.3). But higher education cannot be undertaken solely in 'student contact hours'. There really is far more non-research work to be done in a university science department than in a full-time research establishment, and most of this is work that can only be done by fully trained scientists with experience of research.

A mid-career transition from a combination of teaching and research to a combination of teaching and administration is typical of academic life. Thus, it has been estimated (MacNabb 1981) that 80per cent of academics below the age of 35 participate in research, but that this proportion falls to 25 − 30per cent for those over 50. But this 'attrition' is accompanied by a natural process of 'redeployment' which is much less traumatic than a change of research specialty or of employer. As a research-council scientist said:

'There are other jobs in universities, and faced with that there aren't other jobs here, apart from doing research: that is why [the fossilization of some of the older members of staff here] is a real problem.'

To put this in effective terms: the middle-aged lecturer or senior lecturer who has not been very successful in research can cover up disappointed ambitions by complaining that there is no longer any time for research because of all the teaching and administration he or she is expected to do, and can go on feeling professionally useful in this role until the age of retirement. Such a pattern of self-justification and redirected self-actualization is seldom open to the SSO or PSO in a similar situation in a research council or government establishment.

The successful operation of the traditional academic system does depend, however, on maintaining a fair and sensitive balance between conflicting demands on the time of each person. It is easy to understand why, in the days when jobs were plentiful in all sectors of R&D, many scientists left academia for full-time research posts elsewhere:

'I was getting a bit frustrated with the amount of administration that I was doing as a young lecturer, and also the high degree of teaching that I was involved in . . . . My main interest in research was subordinated to these [activities] . . . . In my first year as a lecturer [I had] 120 students in a [−] class, and you know that was quite a responsibility, and a lot of effort went into them . . . learning to be a teacher.'

Yet those now outside academia 'do feel guilty about not doing any teaching and perhaps I'd quite like to do a little'. Their objection to working in a university is that (cf. §6.3):

' . . . a lecturer, when he takes up his appointment, will have quite a strong [research interest] and be quite an expert. As time goes on you get an administrative burden, and you may go under that. A successful one will have research students who will carry out his research, and he will still be successful through them. An unsuccessful lecturer will be one who succumbs under the administrative burden . . . '.

And those who actually do a little university teaching say that they enjoy it but:

'The people I worked with in [−] University, my colleagues, give me the impression it's not the teaching, it's the administration and the general endless committees of the university life that . . . weren't the case perhaps ten years ago. But now that is the impression I have. They enjoy the teaching, I think.'

Although pressures of this kind have certainly increased in recent years, active researchers in academia still say:

'I like teaching. I think teaching is very important, and more important than many people in the university would like to think. But the teaching plus administration certainly inhibits the research . . . . I have been a full-time researcher. That was very interesting. I think I could do that happily, but equally I do like the teaching a lot.'

They may have felt at first that teaching was 'a bit of an intrusion' on their research, but:

'. . . . I found it more and more intellectually stimulating, as time has gone on, and I have been able to select my courses, and I find that you can get across to the students some of the wider questions of why they are bothering to learn science, why they are bothering to approach research disciplines, and so on. It helps me to keep a wider context in the kind of research that I do.'

It may be, however, that as academics get older, the actual labour of teaching can again be felt as an 'intrusion' on research:

'As time goes by I've probably taught myself the major ways of doing that, and [it's] now a mechanical sort of process. I've been able to, as far as my ability goes, impart information to students, and whether it's one set of students or another I don't think it matters. You still get the same sort of problems, and I use the same sort of approaches . . . . Over the last two or three years I've tried to build up [a more interactive approach]. I now have many more tutorials. I have discussed point with the students very much . . . . [But] I still find it very mechanical. Year after year they are saying virtually the same thing. Even though they are totally different groups of students, the same sort of problems seem to come up. They seem to, year after year, have the same problems, which I try and solve in simiar sorts of ways, despite the teaching methods changing.'

In other words, an academic who goes on for years teaching the same specialized subject, even though this may be much broader than a typical research specialty, can also suffer the symptoms of 'undue persistence' (§5.6). In some university science departments it is customary to rotate teaching assignments so that each staff members is forced to take up and teach a different branch of the subject every few years.

This diversification of teaching assignments is regarded by some academics as a boon:

'. . . . we are fortunate in universities . . . because we can maintain an interest in a subject without necessarily doing research, in that people have teaching commitments. Many of us have to teach subjects about which we are not actively researching'.

The ideal situation', of course, is where

'. . . . you not only teach things which you are interested in, but things which are also going to be of value to the student . . . . You teach badly if you are not particularly interested in what you are doing'.

There is no convincing proof of the widespread belief that university teachers should 'all do research, because it enlightens you and sharpens you for teaching, and you are a better teacher'. Nevertheless, when they are teaching subjects on which they are doing or have done research, most academics would agree that:

'. . . it makes my teaching more interesting, because . . . I can demonstrate an involvement myself with [it] . . . . I may be wrong, but I felt that the information gets across just that little bit better because they knew that I was partly involved with what was going on . . . '.

But diversification in teaching is not merely a valuable means of 'keeping their minds alive'. It can also be a means of acquiring an interest in new research topics:

'. . . universities . . . enable you to move out of your field so much more readily, because questions are posed to you outside your own particular field . . . this is another area where universities are so much more exciting, in that you do get these questions posed, and you are forced to think about issues . . . I can think of a number of new topics that I have researched in which have come to me through that, in fact, because literally I would not have been confronted with that problem had I not been in this kind of environment'.

Generally speaking, university teachers, like the members of some research establishments, live 'in a very pleasant environment' in 'very pleasant places' and some of the older ones get just as 'fossilized'. But it is much easier to arrange things so that they do not 'just sink down to the lowest acceptable work rate'. There is always plenty of useful work to be done. There are usually lively younger colleagues whose own research would benefit if they could be spared some of the burdens of teaching and research. And it is still true that:

'. . . in universities you have the stimulus of the students all the time . . . . It's less likely to happen to somebody working in a university, because they are constantly being questioned about the validity of what they are doing, and they are having to put it laid on the line in a lecture course. [In a research council establishment] we have none of those pressures'.

For these reasons, a move from full-time research outside academia into a university post is one of the best ways of redeploying scientists in mid-career. Indeed, in some R&D organizations, this used to be a routine procedure:

'[It was] quite a reasonable career pattern to work for 10 or 20 years in a government or research council laboratory, and then take a . . . relatively senior university post, which gave you a different range of interests and activities. I's quite a good way of expressing yourselves, having a career.'

This is not to say that a full-time researcher can step into the shoes of a university or polytechnic lecturer without further ado. One of the outmoded traditions of academic life is that all that is needed to teach a scientific subject is to be good at research in that subject. Undergraduate teaching is a professional craft which has to be learnt by hard experience under the critical eye of an expert. Some very able research scientists have not got the temperament for it, and should not be encouraged to take it up. In many cases, also, highly specialized researchers are ill-prepared intellectually to teach elementary courses far back from the research frontier. For this reason, moves of this kind were facilitated when scientists in research units had:

' . . . loose university connections, did a certain amount of teaching in the departments, and of course everybody had the benefit of the university type of environment'.

This has not altogether changed:

'in the current economic situation . . . universities find it easier to ask visiting lecturers to come and give specialist courses, so there is more opportunity for research council people to actually go and do some teaching'.

But this same economic situation means that the universities have practically no regular posts into which scientists from other sectors can be redeployed in mid-career. Indeed, they are having to reduce their own staffs in many departments, even to the extent of trying to declare tenured lecturers and professors 'redundant' (§9.1). The traditional diffusion of very senior scientists into other sectors of the R&D profession still continues, with eminent professors still ending up as research managers in government or industry. But the gates are closed to a flow of less eminent people in the other direction, carrying valuable ideas, techniques and experience into academia, as well as revitalizing themselves by the transition.

What about research scientists being redeployed to make up the shortage of school teachers in some scientific disciplines? They are certainly quite sufficiently familiar with the 'language area' (§4.4) to teach a school subject such as physics, but they lack any of the other skills of the classroom teacher. Like other science graduates, they must therefore go through at least one year's professional training – the Post-Graduate Certificate in Education – before they can be employed in State schools, and then must spend some years gaining experience in junior posts at very low salaries before they are fully established. Many young scientists do enter PGCE courses after successfully completing a PhD and do not regret the move. But for a research scientist in mid-career, school teaching is no longer an obvious alternative profession, and such a move would need to be sustained by an unusual sense of vocation to be more successful than a move into a profession unconnected with science.

7

Fostering flexibility

7.1 Career management

An abrupt involuntary change of research specialty or of organizational role is no longer a rare event in a scientific career. Although most scientists are potentially much more versatile than they think (Chapter 4), many of them are frightened of change (§5.6), and mistrust their ability to adapt to it. Basically, this is a personal problem, which each individual has to face and solve in a different way. It may even be appropriate, in some cases, to 'go into things from the spiritual aspect . . . go into meditation techniques, and that sort of thing', since the problem may well be 'not just our jobs, not just the work and the scientific community', but 'the whole man'. But mid-career change is also a social problem, in that it arises and has to be dealt with in a work environment characterized by relationships with other people. In this and the next chapter we shall consider some of the ways in which such transitions can be made less frightening in prospect, and less damaging in their effects.

Some scientists still resist the idea that deliberate thought should be given to their careers (§5.1). Perhaps they feel that this smacks of 'careerism', which would not be consistent with the image of the scientist dedicated to the pursuit of truth (§6.7). They accept that 'it's important for young people', who 'certainly need for the first few years − ten years, I would have thought − some sort of help and guidance in the system'. But 'people who are in mid-career' are considered to be 'sufficiently mature to set their own targets' and 'judge to what extent they have achieved them'. Otherwise their personal autonomy (§5.5) might be challenged:

> '[Research] is supposed to be a realm of creative endeavour, and as soon as you think about managing the career structure of it you wind up with the Post Office.'

Looking back on 'drifting' research trails (§1.7), they thus tend to emphasize the factor of chance, pointing to the way that 'things have fallen into place, like a jigsaw' so that they have 'been very lucky in being able to pursue the things that have interested them'. This is still perceived to be the typical pattern in many R&D establishments:

'I don't think careers are planned. I think serendipity is a major factor in what happens. In this organization . . . what you do is that you let it be known at the highest level that you are interested in some particular field, and if the opportunity in that field becomes available, as they sometimes do in the changing times, your name is remembered when something comes up . . . . What you have to do is you have to be in the right place at the right time, and the people above have to know exactly what you are interested in . . . . There is supposed to be career management in [this establishment]. I don't reckon that it's more than words on paper. I don't think anything is actually done about it. And I don't think it could be because times change so much.'

The traditional academic or quasi-academic protocols tend to inhibit any outward display of concern by the more senior scientists about the careers of their colleagues:

'Before [the last few years], by and large, the management influence was kind of implicit, and actually quite good. That is, you had quite a lot of freedom about what work you took on, and, by and large, you would be allowed to develop in a particular route . . . . The contrast, apart from your topic area, was applied versus basic . . . . Since the relationship [in the former case] was under continual invention [*sic*] with respect to the particular clients, then you found that people naturally grouped themselves . . . . Management intervened at a rather higher level than that, when, for example, the unit as a whole was getting out of kilter . . . . But they didn't intervene so much at the level of ''You ought to do this'', or ''You ought to do that'', but ''We, as a group, ought to fulfil these commitments''.'

If there are occasions for any discussion of a person's career, they come by chance:

'I am not sure that anyone ever gave me any advice, as it were. What that largely does is very closely related to your current activities. The issue of when you would ever apply for another job, or under what conditions that might be sensible [might be discussed with someone more senior]. Perhaps, in individual cases where people are not satisactory, it's obvious that would inevitably lead to an interaction, and one would want to consider what to do about it. But to forward plan over a longer period of time − ''So what might you be doing in ten years' time?'' − that sort of question − on the whole, I don't think that crops up . . . . While things are going well . . . little is said. Nobody discussed with me, for example, whether I had, say, administrative talents, or that I had a failure of scientific talents even − that really hadn't I considered this, or alternatives, in the longer term?'

The more senior scientists in such establishments do, of course, feel concern about the career problems of their subordinates, but they have little guidance on how to deal with them:

'Your general awareness of what the possibilities are, whether you should interfere in this sort of situation or not — it isn't something which management normally would be exposed to. There's no book you can turn to, and say, well, look, here's the case history one can go by. If one doesn't use one's initiative on this, there isn't that sort of back up . . . . .'

Indeed, in some establishments there seems to be a peculiar reticence about the whole subject. As one of the participants in a group interview remarked:

'It's very interesting, because just sitting around the table . . . I know more about these people — I have been here for ten years — I learnt a lot more about what makes people tick this way than I have in any other way . . . . Maybe one thing that we don't get involved in down here, as much as maybe some other institutes, are these . . . management courses where people go along, and they learn about contacting people, and working with people, and learning how to interact with people.'

This attitude was not seriously challenged until recently, because science was expanding so rapidly:

'So you did not really have to worry about career structure. Nobody thought about it anyway, because it was natural. People would move out . . . . As you go up the pyramid it doesn't get fatter, because you move out and create another pyramid, and that happened in science up to the late 1960s. It's only in the 1970s that we have begun to have the problems.'

The traditional attitude still persists in most universities, and in a few research council establishments. But even in academia the feeling is growing that 'management' must take some active responsibility for career development. This is a particularly serious issue for the large numbers of experienced researchers without academic tenure (§3.3), but it also applies to tenured staff whose promotion is blocked or whose research grants are not renewed. In other words, the traditional 'laissez-faire individualism' of a career in academic science is being increasingly 'collectivized' (§2.1 Ziman 1985), and is tending to conform much more closely to the career of the typical 'QSE' in industrial R&D (§9.4).

For most scientists nowadays, therefore, it is assumed that if research projects are cancelled, or R&D programmes abruptly changed, the management will look after the careers of those affected. For 'people who are not sufficiently self-motivated' to seek a change of field, 'it is up to the management of the organization to do it for them'. In the ideal case:

'The good manager will know whether person X is adaptable or not . . . that is what management is paid for. They are paid to classify individuals — one of the things they are paid to do — in terms of adaptability. So that a good manager looks down his staff, and says, ''Who is adaptable who can I modify?'', and if he gets it right then that person fits into his new role . . . that is the role of the manager.'

What this means, in practice is that managers are expected to watch over the careers of *all* their staff on a regular basis:

'The job of manager is to get the best out of the people under them, and thereby you are giving them job satisfaction. Because it's only when they have that that they are going to be working on maximum capacity. But you've got to size up different people,and try to move to avoid putting the round pegs in the square holes. You've got to try and juggle as far as you can with the staff you are responsible for . . . .'

Unless this is done systematically, however, there is a tendency to concentrate on the extreme cases − the obvious lame ducks and the putative high flyers − and to neglect the moderate majority in between:

'The people who are self-motivated are the people who tend to . . . talk to their management colleagues about where they think they are going, where they would like to go, how the work is developing. People who don't are the people who are probably not motivated . . . . They are waiting to be stimulated from outside − that is, by the management − and that does not happen.'

Senior managers sometimes forget that the traits of highly successful scientists give a false image of the profession as a whole (Lemaine *et al.* 1972). As Bailyn (1980) points out:

'People who do want to move into positions of power, and who are successful in doing so, are the ones who define the rewards in their organizations. With no explicit structures to the contrary, they proceed on the assumption that all high performers have interests and capacities similar to theirs'.'

It is normal now for scientists outside academia to go through some sort of annual interview with more senior staff − a *Job Appraisal Review* (JAR) as it is called officially in the Scientific Civil Service. In some R&D organizations this procedure is undertaken quite informally. For example, the head of a unit tries to encourage personnel 'to consider what their options are' in relation to the programme of the unit:

'All I've done at the moment is − and the contrast is with nothing, so it's nothing very radical − that I make an annual re-appraisal of people, and what they are doing, and what they think they are doing, what I think they are doing, and how it relates to the nature of the activities . . . .'

The head of a university group had instituted what seems a slightly more systematic procedure:

' . . . I have tried to do, as far as I can in a university, a kind of management by objectives situation, where I interview − or at least the Head of

Department interviews — every member of staff, each year, who sets his own objectives and goals in his teaching and in his scholarship, year by year — I think that is the sort of time scale. We review this each year, and see to what extent each member of staff has achieved his objectives, exceeded them, and so on . . . that is a little unusual . . . but I believe, myself, that universities need a much more positive career development . . . .'

'. . . I think it helps both the person in the post to enable him to achieve realistic targets by joint discussions, and then at least a check, at least every year, to see whether he really is on course . . . . It gives some responsibility to the organization. It feels that it is not wasting public money by being sub-optimal the way it is using people . . . . It is a responsibility to society and responsibility to the people whose careers you are trying to develop . . . '.

Even when the procedures are elaborately formalized, a great deal still depends on the personal skills and judgement of the appraiser:

'There is a whole . . . process which in theory . . . is quite good. We do annual reports, via the next man up the line, and then the second man up the line, and the second man up the line also has a chat with the person being reported on . . . . But it depends a great deal on the judgement of these people . . . . Some people are good judges of people, and good judges of what other people ought to be doing, and, just as inportant, good persuaders of those people. Others are not quite so good. They duck the issues, they don't like to be unpleasant, or they think it's unpleasant whereas often it's not really unpleasant, and the chap may well be grateful to hear what he hadn't realized before . . . .'

How effective are such procedures in foreseeing and dealing with career problems? An honest discussion with a sympathetic manager can certainly clear away personal misconceptions:

'One of the problems we've had recently is people who come and talk to you, and say, "But this is the first time anyone has . . . criticized what I am doing. I've always been patted on the back, and told I'm a splendid fellow, and I just need to go on and everything will happen". And you say, "Well, it's just not true. You have been doing some very strange things in your time, and you should have known that." Now people don't take that badly if you're honest with them, and tell them how you see it. You say, "Well, I could be wrong, but this is my view of what you've been doing". They say "But my last division leader always told me, once a year, and said what a splendid chap I was".'

Managerial intentions can also be clarified:

'[The interviewer] actually told me . . . "You may as well forget being a
botanist. When you are old and grey, no doubt you will take it up again. But
until then forget about it".'

The Holdgate Report (1980) strongly recommends 'Purposive Career Planning'
in the Scientific Civil Service:

'While recognizing the importance of personal motivation, and safeguarding
the right of individuals to apply for vacant posts, we are convinced that a
more positive approach by managers to the development of staff careers is
essential.'

In 1981, when the group discussions were recorded, it seemed that the formal JAR
system did have some effect at the local level:

'A lot depends on the guy who is actually giving the JAR interview being able
to make some sort of decisive move in the direction that you wish, and that
he thinks is appropriate.'

But these procedures had not yet been integrated into a system of 'Co-ordinated
Career Planning' of the kind also recommended by Holdgate. Scientists in at least
one establishment 'did not consider it a bad thing if this were to happen', but noted
that there was 'no formal part of the organization called the career development
officer, or something':

' . . . the theory behind the structure is that JAR reports will go down to [−],
and somebody there, who might exist in the future but does not at the
moment, will compile a management system for organizing staff careers −
for saying this person has a potential to do this, or see if this can be organized
in this way. But at the moment the JAR form is filed away somewhere, and
if you are lucky somebody remembers something that you have said on it,
and may even act on it, at some time . . . . The framwework is there, but
the actual management − somebody sitting down there, and saying "God!
Look at this bloke. He is wasted in this job. We will move him" − there is
nobody doing that.'

But here we are beginning to move away from the management of scientists to
much more general issues of the science of management. For example, Hirsch
(private communication) argues that

'The bridge between career planning and systems dealing with individuals
is rarely built . . . . Organizations do not know how the strategic policies
they wish to implement can actually be brought about. Nothing changes in
the way appointments are made, so career problems worsen, and the
management development system drifts ever further from meeting the real
needs of the organization.'

There is much to be said in principle for *centralized* personnel planning as a means
of achieving internal mobility in a large firm, e.g. (Hutt 1981):

'internal vacancies notified centrally, vacancies held until candidates from among surplus staff could be interviewed and considered, where a rejection of such candidates had to be accepted by the centre, where internal promotions were not considered until all likely candidates from the surplus list had been tried, and where recruitment from outside could not happen without the explicit consent of head office'.

But even if such a system did, in fact, lower some of the *bureaucratic* obstacles to mobility for scientists in mid-career, it would not deal with other aspects of the problem. As we shall see, R&D managers have a number of other means of helping their staff adapt to change, but they cannot take all the responsibility for their careers:

'I learnt a lot of things from my period in personnel . . . . You get some people who come in and who believe the company will do the right thing for them. All they have to do is what their managers suggest, and it's bound to be in their own best interest as well. And the managers say, "I want you here", and you stay here: another manager says, "You do six months there", and you go there, and so on. And you get other people who are normally labelled as Bolshie, who have their own feelings, views, on how to follow their own careers, and it's very difficult if you have clear ideas on what you want to do within [an organization] — it's very difficult to find the opportunities to do that if your ideas differ from those of your management.'

7.2 Training and recruitment

Careers are being managed before they even begin! The way in which scientists are educated, trained and recruited into research must surely have some influence on their later reactions to change. Should educational and recruitment policies and practices be modified, in the hope of making scientists more adaptable in mid-career?

It is often asserted that research scientists owe some of their narrowness to the specialized courses they took as undergraduates. But as we saw in £4.3, Special Honours degree courses are much broader in scope than the research specialties into which scientists often settle. By the time a scientist has reached the age of 40, the subjects that he or she may have studied at university are no more than a sketchy landscape whose details have been largely forgotten, or else have been redrawn with much greater precision as a result of more recent and compelling work experience.

It is doubtful whether scientists can be trained to 'be more adaptable' by formal courses of study at a university or polytechnic. Nevertheless, broader degree courses must surely be desirable as a means of introducing students to interdisciplinary subjects. 'The sort of approach that the training gives them', makes them 'quite happy . . . to take the subject as a whole', and thus appreciate from the start the full extent of the scientific 'language area' (§4.2) over which they may eventually roam. But this approach may be a disadvantage when it comes to getting a research job:

' . . . the course selects the kind of person who has not yet decided that he wants to specialize in anything. He wants to keep his options open, but to continue in higher education . . . . A lot of people come to the end of that degree course still not knowing exactly what they want to do in life, and certainly the students we have had [on sandwich courses of an inter-disciplinary character], which are now six or seven, all more or less fall in that category'.

'They have a very broad education over a wide range, and they still haven't − most of them − come to the point where they have decided what to do in life . . . . These courses tend to attract people who have not made up their minds, and perhaps lack the capacity to make up their minds, but who, in the process, get a very broad general education which perhaps some of us lack . . . . But they lack the specialist knowledge to work in an [establishment] like this [unless] they do firm up on some specific field and then do a postgraduate course on it . . . .'

' . . . From the ones we have seen, which are the cream of the students [from a well-known university] . . . , most of them are still a bit vague at this stage. Some of them have matured while they have been with us.'

' . . . One of the sad things is, we have very little opportunity to employ them if they finish their course . . . No way at all do they satisfy the criteria for employment here as graduates.'

Traditions of specialization in academic science and narrowly specialized criteria for short-term employment can thus defeat the effort to give a broad education for a lifetime of change.

Undergraduate education in the natural sciences is only indirectly vocational. Postgraduate training in research is much more formative. Graduate students are forced to become very narrowly specialized in their scientific interests. Even though a student may have come into his or her particular specialty without serious consideration of alternatives − 'it was the line of least resistance . . . I had a very good department where I was . . . I stayed and did . . . a good solid type PhD' − these interests are often retained for a lifetime (§1.4). At the same time, doing a PhD brings the student up against the snags and frustrations of research. The experience of having to 'stick with it', and 'see it through' makes them 'unlikely later in life to be easily persuaded just to give up'. The psychological foundations are thus established for undue persistence in a narrow field of research (§5.6) and extreme resistance to change.

During this period, the student also becomes acquainted with the academic style of research (Lemaine *et al.* 1972). They work under, and to some extent alongside, academic researchers. They take part in seminars and conferences, and may even become the proud author or co-author of genuine scientific papers. They thus internalize many of the unwritten norms of the scientific community (Ziman 1985), get a taste for 'recognition' as a researcher in a particular field, and find themselves already standing at the foot of the ladder of a 'reputational' scientific career (§3.3, §5.4). In other words, their training in research does not prepare them in any way for the 'organizational' careers that most of them must in due course follow, and does not provide them with a wider range of technical skills than the minimum required for their actual research.

It is not surprising, therefore, that industrial firms are in two minds about hiring PhDs. On the one hand, they bring with them specific expertise, intellectual talents, and experience of working away at a problem on their own. On the other hand, they 'want a job along the line of what [they] now feel is [their] special-ization', and it may take some time for them to adapt to the atmosphere of urgency and managerial authority of industrial R&D, and to catch up with their con-temporaries who came into the firm straight after a first degree.

Graduate students are also doubtful about whether they ought to go into industrial R&D. As they approach the end of their PhD course, many are still attracted by the idea of 'staying on' in academia, despite its competitiveness, insecurity, and heavy burdens of teaching and administration (§6.5). After all, this is the life they know, with a familiar career structure and role models to emulate or transcend. There is little informed discussion of the place of the PhD scientist in other sectors of the R&D system and they have seldom had the occa-sion to take part in any other forms of scientific work.

They thus enter the search for employment with limited information and very uncertain motives:

'I did not know anything about [this establishment] until I got there, and decided it would be quite a good thing. What I wanted to do was to try and get into some job which had got some scientific content, but also had some applied side to it. I did not want a purely academic type of job: I had decided I was not really cut out for that. So I got the job by sheer fluke. And, like everybody else [here], I am working in a totally different field to what I was trained in.'

Even when there was no difficulty in getting a good job they were not always very choosy. If a major industrial firm where it would have been possible to 'remain in the field' was 'not recruiting that year', then a job in an entirely different field in the Scientific Civil Service was acceptable. Other factors often weighed more heavily than the subject of the research:

> 'I guess what I wanted to do at that time was something practical. I wanted to see it — my research — have some practical benefits for the population, rather than another [piece of academic knowledge]. However, I felt that the whole Civil Service situation just seemed to me to be too cushy. It was too easy — it was, too, you know — and I guess they weren't going to pay me enough money. They . . . gave me an extra £150 for my PhD, and I thought there's more I can do with that. So, in spite of the fact that I hadn't got a fellowship anywhere, I rejected that job, and I thought actually that was quite a difficult decision to make at that time — quite a courageous one.'

That was in the 1960s: yet the Holdgate Report (§1980) still has to insist that:

> 'Neither the diversity of work done by scientists in Government nor the range of career openings is sufficiently well known and we *recommend* that the Civil Service Commission re-examine their recruitment literature and that more is done by departments to publicize the work of their scientists.'

The whole process is too haphazard to support the notion that the scientists entering any particular R&D organization are self-selected for the type of career they are likely to lead in that organization. Nor would it seem that organizational recruitment policies always have this purpose in mind. There was a time, indeed, when the whole emphasis seemed to be on strictly 'academic' qualities. In a quasi-academic research council establishment it would be natural enough even now to say that:

> . . . 'the major thing which matters in a fundamental research atmosphere — . . . the only thing that matters is ability . . . ability to define a problem and to carry that problem out successfully'.

The director of an elite establishment in a fundamental interdisciplinary field reported that it is still his policy to recruit PhDs from *any* subject solely on the basis of 'how bright they are'. Twenty years ago, a government R&D establishment mainly involved in urgent technological development would also have been recruiting people on this basis. A recent PhD might come as a 'senior research fellow' into 'a team of ten PhDs . . . virtually a university department':

> 'I was amazed that they'd offered me a job without saying what it was about, and so I wrote back saying that I want to do [subject X], because at that time I thought that was the only thing worth doing, and so they said, ''OK, then do [subject X]''. Apart from that, they didn't seem to attach any importance to what you did: the only thing they expected was to roll up on a Monday morning.'

Even industrial companies were recruiting PhDs to go into fundamental R&D because:

. . . 'it gave them an opening into academic circles, it provided people in the company who actually knew what was going on through participation in academic science, and therefore to transfer these ideas better.'

In practice, of course, it was not really true that the management of an R&D establishment:

' . . . did not know which direction we were going in, and we were recruiting and letting him do what he likes. It was not quite like that, although you might get this impression . . . . He was allowed to develop his expertise. We recruited him because he had the expertise, and he seemed to be the sort of man who would develop that expertise in the way at that stage we thought he was going . . . . What we were looking for when we got him was somebody we felt . . . could develop in that particular aspect of the work . . . . OK, he has put his own imprint on it, but . . . he was recruited to do that. It's not really quite as haphazard as one might have suspected.'

But even a graduate, without a PhD, entered the civil service as scientist 'whose role was going to be developed in the future', and would go through:

' . . . almost his "apprenticeship" period, where . . . as a Scientific Officer, you knocked around a few different departments, you did a few different jobs, until either you or the system found somewhere where you really slotted'.

The long-term career implications of such policies were obvious. They conveyed a clear message to the new recruit. He or she was being cast in the role of the 'honest seeker after truth' (§9.3), who will be expected to take the initiative, not only in the performance of research but also in pursuing a personal career within the organization. From the very start, the management was tacitly telling its employees that it would provide them with a stable organizational environment but was not taking any responsibility for their ultimate fate.

At a time when science was expanding, and jobs were permanent, such an attitude was acceptable. It offered opportunities for the venturesome and security for the cautious. But it is unacceptable in the present atmosphere of retrenchment and redeployment (§2.5, §9.1), where justifiable ambitions are dashed and professional devotion is rewarded with redundancy, through circumstances beyond any one person's control.

The new management structures that are evolving to meet this situation call for new recruiting policies and practices. These policies have certainly changed a lot in recent years, but in quite the wrong direction. The general effect of financial constraint has been to put all the stress on the acquisition of staff with immediately useful technical expertise:

'In the present economic climate, the only way we stand any chance of actually recruiting anybody is to have a fairly tight job description . . . so there is little opportunity of recruiting a good person to the [establishment]'

' . . . you can take it further. There has got to be an external body available, and even then it is going to be difficult. You've got to be able to say that there is nobody available who could do the work instead'.

In many government and research council establishments, it is no longer possible to take somebody with a broad first degree, or with experience in several fields, or with 'a whole area of new skills' and 'mould them into a job'. Scientists with relevant specialized skills are recruited directly into particular research programmes, rather than being chosen to fill a broad gap in a 'multidisciplinary team'. Because of the competition for jobs, highly qualified research scientists are accepting routine technical posts in commissioned research projects. The balance between the 'scientific' and 'experimental' functions is being upset:

'The establishment has a policy of appointing, as far as possible, first class Honours graduates, almost to the total exclusion of what we would call 'old style assistant experimental officers' − that is, people who've done a practical training course in electronics or workshop metalwork, and so on.'

R&D managers know very well that attempts to match recruits against the immediate needs for specialists are just as misguided as the assumption that the changing 'manpower needs' of industry can be provided by matching courses of Higher Education (Hirsh 1982). Their desire to increase the mobility of scientists in mid-career by making outside appointments at a more senior level (Holdgate, 1980), is frustrated by the need to appoint people as young as possible in order to level up the age distribution in the scientific community (§3.5). They see short-term temporary appointments as advantageous to their organization, but very un-satisfactory for those who must take them (Holdgate 1980).

But the primary objection to present-day recruiting practices is that they reinforce the role of the scientist as a specialized 'expert', rather than as a generalized 'problem solver' (§9.3). They are thus quite inconsistent with the primary need of every R&D organization for 'adaptable, flexible' people (Hutt 1981). Fiske (1979) quotes a well-informed respondent:

'The quality which I have observed to be most important for obtaining a job in the aerospace industry is highly specialized expertise in the precise area the employer is looking for. The qualities which I have observed to be most important for long-term job satisfaction and for holding a job in times of decreasing funding are flexibility of outlook and breadth of knowledge in several areas of physics.'

7.3 Early diversification

Scientists who have experienced changes of subject early in their careers find it much easier to adapt to change later on. They have more skills at their command (§4.4) and are more confident of success (§5.5) than those who have stayed all the time in the same specialty. Although this precept was seldom stated explicitly in the group discussions, it is clear that most people with long experience of R&D organizations agree that 'it's useful to give almost everyone varied experience in their early years', and that 'when people are young they do need a push . . . to get them into a new area where "there is this problem, I don't know anything about it, will I be able to cope?", and so on.'

It is also generally agreed, however, that it takes many years of undivided attention to become the real master of a research specialty (§5.3). On the face of it, these two precepts are mutually incompatible. Nevertheless, they can be reconciled by encouraging a certain amount of diversification (§1.5) at the early stages of a research career, followed by a period of concentration on a single theme:

' . . . there is a need to push people – not to take their reservations too seriously – when they are young. But . . . it is important to do it when they are young . . . . When they are older – and it varies from person to person – they should then be settling down into an area that both suits them best and suits the company best. Otherwise people are moving all the time, and never achieve sufficient knowledge or experience to ever contribute anything'.

This strategy is not, of course, consonant with the erroneous notion that all scientists do their best work when they are still very young (cf. §4.8). This notion is only valid for one or two disciplines, such as pure mathematics and theoretical physics. Generally speaking, research scientists do not reach their peak of productivity until they are well into their 30s. It turns out, for example, that 39 is the average age at which outstanding scientists actually did the research that later won them a Nobel Prize (Zuckerman 1977).

In fact, many scientists do go through such a period of diversification in their late 20s and early 30s. A significant proportion of those who took part in the group discussions reported that after a first degree or a PhD they had had two or three quite different research jobs, in different R&D organizations, before settling down to a permanent post in their present establishment. Although these jobs were sometimes in quite different subdisciplines (§1.2), they mostly involved quite serious and specialized research, and usually lasted for several years. In other words, the early careers of many researchers who are now in mid-career had actually followed the pattern recommended by Pelz & Andrews (1976):

'Up to age 34, devoting two or three years to one undertaking is a source of security. It enables the young man to build contributions in which he can take pride. But it is still better not to be too pre-occupied with digging deeply.'

In many cases, they had spent several years working abroad, not only as researchers but also as technical workers for commercial firms such as oil companies. They might also have moved through several sectors of the R&D system – from a university research assistantship, say, into an industrial firm, and then into a government laboratory. Early diversification in search of employment has always been a significant source of versatility and adaptability in the research profession.

The striking increase in the proportion of researchers in short-term appointments (§3.5) has undoubtedly enhanced this factor in recent years. It is true that these appointments are usually advertised for a person who is already highly specialized and ready to carry out research on a prescribed project, so that a researcher who passes through a series of temporary appointments may actually be working all the time in the same narrow specialty. Nevertheless, the matching of projects and expertise in the scientific job market is very imperfect, and many highly competent and well-motivated scientists are now forced to move from job to job, often having to discard their existing expertise and acquire quite new skills at each move.

Up to a point, this involuntary mobility does generate versatile and resilient scientific workers, willing to take on any R&D problem offered to them. But the quality of the work they are capable of doing will suffer if successive projects are too short, or if the process of diversification goes on for too long in any one person's career (§5.4). Clever and self-confident 'trouble shooters' are certainly of great value at all stages of the R&D process, but the 'jack of all trades' (§4.4) can never completely replace the expert with ten years of concentrated experience in the relevant field.

But a change of specialty does not necessarily require a change of employer. Many mid-career scientists reported that after joining their present establishment they had worked on a variety of different projects, involving a variety of research specialties. In some cases they had moved temporarily into an administrative or service function (§6.3), or had moved out of the research laboratory into an industrial production unit for a few years:

'[When I came to this firm after taking a BSc] I focussed my attention on being a [–] technologist, to stay in R&D to become a specialist. After being here something like two years I was posted down to one of the works and a production problem that was to last "of the order of three weeks" – and I reappeared 18 months later. That was a great change of experience for me, in the sense that I actually worked on plant on a problem, very early in my career, not knowing a thing about the problem I was going to work on, and that opening really led me naturally to look at the works environment of R&D, and working closer to a proces. From there I joined a different group . . . which was a research and development back up service to [a branch of the firm's operations]. I went into the [−] division initially for a short period, and then I went on a course [abroad] . . . and came back into [another division] for a short spell − and then into the works again as a representative [of the R&D service group]. At the end of [a] four-year period I was offered the same post in a different works . . . [using] a totally different process . . . . I spent some three years there, and I was offered a production post back in the [original] works . . . .'

Early diversification is quite normal for scientists in R&D organizations where 'urgent' projects (§2.3) come and go rapidly. Many R&D managements also have a formal policy of encouraging internal mobility for new recruits. For example, a government establishment used to have a scheme 'by which all new entrant graduates spent some time in approximately one year in three different departments, just to give us some feel for the establishment'. New graduates coming into an industrial company were reviewed for the first three years by personnel managers 'at least twice a year, discussing both with them and their managers, or senior management, what they had been doing, how they had been doing it, what would be best for them next, whether they were ripe for a move, if so where'.

But this induction period is usually quite short, and for those who come straight from a first degree may be over by the age of 25 or so. Once the new recruit has 'settled in', he or she is often left in the same job for many years. Although managements say that they favour internal mobility, they also appreciate the costs and disadvantages of excessive diversification (§5.4), and tend not to move people except to meet the immediate needs of the organization. Even if they are watching over the careers of their staff, very few R&D organizations seem to have systematic procedures for moving people from specialty to specialty over longer periods. In some, 'the top administration don't like the idea of mobility'. In others:

A. 'There is no formal way . . . in which people do change their fields within the establishment. It normally happens because something has closed down, or for some other reason, or whatever. But there is a general guideline sometimes that people are persuaded to move areas.'

B. 'I don't think that's true actually. I think there is a policy, if you can call it that, of people moving. In fact the Director has stated it twice in the . . . Annual Reports.' . . .

A. ' . . . But there's no policy on how you actually do that, and deal with the personal − and retraining in the subject.'

The difficulty about any such policy is how to carry it out in harmony with the wishes of the individuals involved. In one particular organization with a tradition of moving people arbitrarily at short notice, every few years, this is now seen to be essential. Although such moves are still officially planned in secret from above, they actually involve 'putting options' or consulting 'the grapevine' about where 'people who are due to move in a few months' time, or express a desire to move' would like to go. In practice:

'The [senior management] are keen to have suggestions to work on, and that almost certainly implies that the characters that are most forthright and forceful are likely to be satisfied, and the others are not quite so satisfied.'

This particular organization is unusual in that it does manage to combine a high level of R&D whilst enforcing mobility and diversification on its staff. This would certainly not be favoured in many other R&D organizations, where:

' . . . we would prefer to decide ourselves, as far as possible, on mobility, rather than having a system whereby you are moved from A to B without really having been consulted . . . . Many years ago, when I was quite young, [I remember] being rather incensed by a very senior group of people . . . who came and said . . . they thought it was an excellent thing for people to move around every five years, and so forth. They had no intention of moving around themselves. And it was at a time when would have been very inconvenient for me. I think this needs a great deal of sensitivity and understanding by management'.

As a consequence, although scientists at all levels feel that 'we don't have enough mobility, and mobility is something which has to be forced on us', they are very reluctant to put this principle into practice. In most large R&D organizations, the only systematic administrative facility for internal mobility is a system of internal advertising of vacant posts, i.e. 'trawling' for suitable candidates throughout the organization. These procedures are preferred by the unions because they are equitable in principle, and open up opportunities for people whose desire to move has not been recognized. But they are generally considered to be ineffective as a means of achieving the level of voluntary redeployment that is now considered desirable.

It is obviously difficult to encourage early diversification for a scientist working in a small, geographically isolated establishment, because of all the costs and disadvantages of moving house (§5.8). Nevertheless, many geographically dispersed organizations have developed strong traditions of mobility from unit to unit (Hutt 1981). Financial and administrative factors also tend to constrain such moves to the minimum required to fill a vacant post or give employment to someone whose previous project has come to an end. But an active policy of internal mobility can also produce considerable administrative benefits. Thus, as Hutt also points out, some firms have arranged:

'a chain of related moves, in which each employee was moved to a job requiring slightly different skills from his original one. A surplus of physicists and a shortage of software people could, and sometimes was, solved by a chain of moves from physics to electronics, from electronics to computer hardware, and from hardware to software. Similar possible chains of moves could be identified at many level and of many kinds.'

In the long run, formal managerial policies and administrative practices are less important than the development of an 'atmosphere' of good will and confidence within the organization. There can be little encouragement to move around in an establishment where 'people should have been co-operating with each other, and had not, and had been keeping themselves apart' or where 'there is not a great deal of cross collaboration between various departments, various groups, and there is one seminar, once a week, which is given mostly by people outside the lab'. In another establishment:

'the project is standing still and treading water . . . . If any of us had a good offer from somewhere else where you could go and do some real active work, I think most people would go'.

Yet they stay put, immobilized in their disaffection, simply because they have no tradition of active management for diversity and change.

7.4 Team work

Scientific work used always to be organized in terms of subject specialties (§1.1). In a university, the physicists are gathered together in a physics department to work on physics problems, the chemists work together in the chemistry department, the bacteriologists in the bacteriology department, and so on. Although academic scientists have always been allowed a great deal of freedom in their choice of research problems (Ziman 1985) the administrative barriers between department and department have always been difficult to cross. Over the years, the map of knowledge slowly changes (§1.2), but for the scientists developing a new subject it may take years to establish a new interdisciplinary department – geophysics, molecular biology, marine ecology, etc. (Lemaine *et al.* 1976). Most scientists, in fact, do not 'actually . . . move out of' the 'little box with a label above it' where they started their research. Unless academic departments are very large, their boundaries are major organizational constraints on intellectual mobility and adaptability to change.

Some R&D establishments are still compartmentalized into subject specialties – geologists (say) in one section, botanists (for example) in another, mathematicians (as it might be) in another. In such an establishment, 'multidisciplinary' projects, drawing expertise from several sections, are not encouraged because a section might 'lose a staff member that they could not replace'. Even in 'areas where the people have offered to collaborate in a very friendly, loose way . . . at the end of the day they've . . . said "Oh, I'm sorry, I couldn't manage that . . . the boss wanted me to do something else".' Sections are subdivided and multiplied to provide new units of management for newly promoted staff (§6.2). In other words, the whole organization of the work of the establishment seems designed to reinforce narrow and rigid specialization of subject and role.

But as R&D projects have become more relevant, more urgent, and more extensive (§2.3), they have necessitated 'group efforts, bringing together different skills, to tackle them'. This tendency can be observed in the majority of R&D organizations outside of academia:

'we are now, more than ever, being encouraged not to work as individuals but to work in groups. It is now . . . accepted [in this R&D organization] that we all work in groups, [and] that there are fewer projects and [that we] put more effort into these. That is the way things are really going at [this establishment], now, in organization terms. It means . . . that people will have much less chance of being highly specialized than they were before'.

Quite new administrative structures have had to be devised to manage *team* research, which inevitably cuts across the traditional boundaries between specialties. Very extensive technological development projects generate elaborate managerial structures, sometimes stretching over a number of R&D establishments, to orchestrate all the skills and resources they require. In simpler situations, the teams themselves are treated as primary administrative units, even though they are not permanent:

' . . . we have quite a large series of "laboratories", so-called, which are groups . . . built around people . . . . Should those people leave — and, indeed, once or twice people have left — the thing has been closed down as a unit . . . . It has the advantage over the ordinary [self-contained, quasi-academic] unit that when a laboratory closes . . . one can, hopefully, very easily fit in the people within the organization itself . . . '.

Many organizations nowadays have set up some form of *matrix* structure, where each project is a recognizable, if temporary, managerial entity, but where the members of the project team retain their long-term attachment to their respective subject or professional divisions. It would take us too far afield to discuss all the pros and cons of 'matrix management' in R&D (see, e.g. Thomason 1970, Knight 1976, Gunz & Pearson 1977). But from our present point of view it is obvious that it must greatly facilitate mobility within each establishment:

'The large "tribes", which are the [line management] divisions, provide a sort of stable basis for these people, to live and feel comfortable [and] are formally responsible for their careers and so on . . . . This greatly aids the transfer of people, via the project system, from one type of work to another, because they have this home base, [where] they feel comfortable and stable, which provides their basic family, and therefore they are happy to switch the nature of their work through the project system . . . . If we tried to switch them from division to division each time their work changed . . . we would have many more problems.'

A matrix structure also enlarges the knowledge and experience of employees by forcing them to deal with subjects outside their specialty, by involving them with decision-making that would otherwise be done higher up, and by getting them known in wider circles (Knight 1976). Zeldenrust & de Laat (1982) found that both 'middle echelon' and 'lower echelon' people developed a better view of the organization as a whole through working on multidisciplinary projects. Hutt (1981) similarly noted that the members of a team drawn from five different units to undertake a particular project had acquired a better appreciation of the benefits of moving between units on a more permanent basis. In other words, whatever may be its disadvantages in other ways, 'matrix management' is an effective means of making both individuals and institutions more flexible and adaptable to change.

Multidisciplinary team research is rapidly becoming the normal mode of operation in most R&D establishments because that is the best way of tackling the 'urgent' and 'relevant' problems they are now expected to solve. Nevertheless, the introduction of this new mode of operation has met with some opposition because it threatens the 'reputational' careers of scientists as subject specialists (§5.3). How is it possible, for example, to build up a personal reputation when the publications and recognition are all in the name of the team as a whole, or its achievements are attributed primarily to its formal leader.

In principle, as a result of an administrative transition from voluntary collaboration to formal team research:

'You can let more people in, because in a group you can have a gatekeeper, who holds open the gate, and says, ''Would you like to come in, because we've got a particular thing you can do?'' That wasn't possible for individuals.'

In practice, however, an active team 'can't carry passengers'. Life becomes very difficult indeed for those who are 'not successful' in the team and are 'returned to their original disciplines', even though they may be perfectly competent as individuals.

Again, if a team is very large, and lasts for a long time, it can become yet another rigidifying structure, where each member plays a narrow and specialized role. At the other extreme, a small group can become 'almost too cosy' if 'everything is going very nicely' for a long time. In any case, because the purpose of the team is to carry out a particular project, the career development of those involved is given a low priority. It is notorious, for example, that project managers are extremely reluctant to release their ablest young people for other work, even though such a move would be best for their careers. A team may be broken up, or its programme completely disrupted, if senior management insists on pulling out a few key person for that sort of reason.

The career of the team thus entrains the careers of its members:

> 'One way of making your mark [is] if one is associated with a project you may have been just transferred to, which then expands, and is seen as important. It has a drive and priority associated with it. There are oppor-tunities in that situation . . . .'

But there are also occasions (§5.6) when people's careers mark time for several years as a project is 'running down', the team structure is 'crumbling', people are getting 'fed up and frustrated', and gradually leaving.

7.5 Promotion

The procedures for promotion have a significant influence on the attitudes of scientists towards their work. People know very well that the decision will depend on some sort of assessment of their performance over a period of years (§7.1), but in some organizations they are often quite uncertain how this assessment is made:

> 'It almost seems to happen by magic, in that there is not a lot of dialogue between the people who are making the decision about who should be promoted, and the people themselves . . . . For example, . . . I knew nothing about it at all, until one day somebody told me I had been promoted to PSO . . . . There was no dialogue whatsoever . . . . The criteria by which people decide who is ripe for promotion . . . should be a bit more widely circulated . . . . It is a recognition [of] what people think about your work, and we get very little feedback about how these decisions are made.'

This discreet veil over promotion procedures and criteria was acceptable in the past, when most R&D establishments were very stable organizations with settled conventions of management. In academic and quasi-academic institutions, it was clearly understood that promotions were decided on the basis of 'published con-tributions to knowledge'. It was taken as a matter of principle that a scientist's organizational career should closely parallel his or her reputational career (§3.3), as judged by expert external authorities.

In practice, of course, considerations such as managerial competence would also enter into the decision. Nevertheless, in many research council establishments

— and even in some government R&D establishments — almost all appointments and promotions up to a certain level still follow the academic model, where everything is made to depend on the length and strength of the candidate's flow of publications. As a consequence, researchers are very reluctant to make any change of specialty or organizational function that would interrupt this flow (§5.3).

On the other hand, in a typical industrial R&D organization the criteria for promotion are much more diverse. A scientist naturally expects to be judged by the technical quality of his or her work, as assessed by the management, but would also expect them to take into account evidence of other personal capabilities such as originality, conscientiousness, maturity of judgement, capacity for leadership, loyalty to the firm, etc. Where the research is of the kind that can be published in the open literature, an impressive list of papers would bear witness to technical achievement, but would not, of itself, be the decisive factor. In other words, the criteria for promotion relate to the actual or prospective role of the person in the organization, regardless of whether this is or is not narrowly specialized.

The economic and administrative constraints of the past decade have pushed most research council and government R&D establishments towards much more 'urgent' and 'relevant' work (§2.5). This, in turn, is putting traditional promotion procedures under severe pressure. A major complaint about commission research (§2.4), for example, is that 'it is not always possible to publish the information as quickly as you would like', so that:

'. . . a worry for a lot of us is . . . that we have grown up in a system which judges the progress of a scientist in a career line by publications, yet we are working in times when the opportunities to publish are much reduced. We are involved with contracts with industry . . . [which] it is important not to disclose. All these things tend to reduce the opportunity to publish.'

Commission work also reduces the opportunity for people to show their capabilities as 'scientists' (§3.4, §5.5):

'I could [get to the point] where the only future for me is to go out of government science and take a job, say, with a consultant engineer . . . . I would certainly not score any points at all [for promotion] if I tried to mix in with people who work entirely with 'science' type of research . . . . I am serving often the requirements of whoever lets the contract that they take on. Their objectives may be non-scientific. I am getting scientific fall-out as I go, but their ptime purpose in using me in the establishment is [for an industrial development project] . . . and that does not equip me well to compete for senior scientists' jobs in the Civil Service . . . . I am sort of a second-class scientist here, I suppose, because of the nature of what . . . I have been trying to do.'

Again, technical support services (§6.4) are growing in scope, yet:

'. . . the support workers, who were having to face a more rapid technological development than the scientists themselves are getting decreasing credit for the contribution they make to the scientists' research. We are now at the stage where frequently a major input has come from the support workers. They don't get publications, they frequently don't get acknowledgements, and the trend for promotions these days is to go solely on the strength of scientific publications, even for support workers . . . . This is a big disincentive to the support workers in scientific research now'.

Team research (§7.4) presents another characteristic problem:

'The climate has changed . . . now towards group effort in most of the projects we tackle . . . . We'd be hard put to it to show hard evidence that [our organization] necessarily would recognize all these joint publications in the same way that one could look at the contribution of individuals . . . . I remember one scientific advisor once sayiny to me, with a smile on his face, "Well, you know, we din't actually weigh these publications!". But there was something towards that about it, and . . . up to PSO that was very much the thing once — to go for single publications and get as many [as possible].'

Even in academia, staff are now being assessed by staff—student panels on their effectiveness as teachers (§6.5), and are also having to show that they can bring research grants to the department, in addition to giving evidence of their standing in a particular research specialty.

Throughout the research council and government sectors of R&D, a serious inconsistency is thus developing between promotion procedures and work roles:

'On the one hand you have this industrial type of work, where . . . you have to do the job on time, you have to meet the specifications, and on the other hand we are largely controlled by academics, who are using academic criteria to assess your work . . . . In the same organization you've got different ways of assessing performance.'

Middle managers are uncertain how to advise their staff:

'. . . there is so much controversy, because people are doing things, and asking whether that will get them promotion — whether they should have lots of publications, or should they stress the payment for the [commissioned] work they are doing — and we can't give them the answer because . . . we don't know at what level the criteria are being made for those promotions . . . .'.

This is not to suggest that these establishments should give up basic scientific research, or discourage their staff from publishing as much of their work as possible in the open literature. Even industrial R&D workers know the importance of 'exposing their work to other people':

' . . . While you are still publishing, you feel . . . that you are still available outside, you still have got a record that other people can look at. Once you stop publishing, there's a decision that says "Well, maybe I may not be able to talk about this sort of work to other people. How can you assess what I have done — how good I am?".'

Indeed, they feel that their job mobility depends on being able to 'hang on to their wares':

'If you can go along and drop a pile of publications on the desk when you go for interview, then you have got something very concrete to offer to people . . . . In times of redundancies and contraction . . . people would actually feel more strongly, if anything, about giving away . . . that.'

It is clear, nevertheless, that there needs to be a major change in promotion procedures and criteria in governmental and quasi-governmental R&D establishments, playing down the traditional virtues of narrow and rigid specialization and rewarding work displaying technical versatility and organizational initiative. Some scientists now realize that something more than lip service to adaptability is called for:

' . . . in practice, the image is not there, and there are still many people who do play the system very much as in the old days in the senior ranks and on promotion panels . . . . I would like to see more questions asked . . . like. "Have you expanded your area of work?", or almost "Have you changed it?" — not as a sole criterion but as one aspect to chalk up as a sort of white one, rather than as a black one — and "What efforts have you made to collaborate with our other people in related disciplines?" and so on: and people [becoming] a little worried if they can't answer in the positive, and a few examples of people promoted in these areas . . . '.

7.6 The role of the senior scientist

In the Scientific Civil Service, the grade of Principal Scientific Officer — PSO is supposed to be the *career* grade — the grade that a well-qualified scientist can expect to reach in his or her 40s, if not before (§3.3). Promotion up to this level — or to the corresponding level in other R&D organizations — depends primarily on competent performance. It is not directly related to a particular post or organizational role. Thus (Holdgate 1980), some PSOs are active researchers, others lead research groups or manage project teams (§7.4), others carry out administrative (§6.3) or service (§6.4) functions, and so on.

The majority of PSOs who are already in mid-career are unlikely to be promoted to a higher grade, but will continue to carry out functions of this kind for the remainder of their careers. This does not mean that their further contributions to science and technology should be discounted. They are expert and experienced scientific and technical workers, often much more adaptable than they seem. In the next chapter we shall consider the various means that can be used, where necessary, to exploit their latent versatility (Chapter 4) and help them to move into new fields of research (Chapter 5), or new organizational roles (Chapter 6).

Most posts above the career grade are managerial or administrative (§6.1). Personal achievement as a research specialist may play an important part initially in winning promotion to such a post (§7.5), but other, more general capabilities are called into play thereafter (§6.2). But a small proportion of the senior staff in some large R&D organizations are what might be termed *senior scientists*: they are ranked high in status, but do not carry the corresponding managerial or administrative responsibilities (§3.3, §6.2). Thus, for example, in the Scientific Civil Service (Holdgate 1980), about 15per cent of SPSOs and DCSOs hold their posts by 'Individual Merit Promotion' − a proportion that rises to 25 − 30per cent in some of the research councils. Some of the major industrial R&D organizations have *dual ladder* career schemes of a similar kind.

These schemes draw their inspiration from academia. In most British universities, readers and professors are appointed on their public scholarly record as judged by independent external assessors. The title of reader is not usually attached to a particular post, but is purely personal. A professor may have been selected from inside or outside the university to fill an established 'chair' − not necessarily coupled with headship of a department − or he or she may have been promoted internally to a 'personal chair' without formal departmental responsibilities.

The primary duties of senior academics are not directly managerial. They are appointed in the expectation that they will make further contributions to knowledge, through their own research, through their teaching, or through the organization and leadership of the work of others. Many do get completely bogged down in administration (§6.5), but others continue to be active researchers, working on their own or with small groups of colleagues and students.

The IMP scheme follows the academic model very closely. Candidates are judged almost entirely on their research publications (or, in classified work, by comparable contributions to the science of their technology). A DSc is one of the conventional criteria (§5.3). External assessors are told that the standard is the same as for a readership or chair in a good British university. It is obligatory to report that 'Dr X is already a leading international authority on the subject'. A hint of the possibility of forthcoming election to the Royal Society is of considerable help − and so on.

Promotion for 'individual merit' thus extends into the higher grades the type of quasi-academic career path that is already favoured for promotion to the grade of PSO (§7.5). By applying the same criteria more rigorously and selectively it reinforces the same tendencies towards narrow and persistent specialization (§5.3). Scientists who think that they might be in line for such promotion are thus deterred from making a change of specialty that would 'significantly reduce their chances' in this highly competitive situation.

Scientists themselves are in two minds about the value of IMP. On the one hand, they see it as:

'the only possible stimulus for a PSO to really do any more work after he has become a PSO . . . unless you have an interest in managerial responsibility, which not everybody does . . . . People have aptitudes and interests . . . . Mine don't happen to lie in the field of management'

They naturally like the idea that one can:

'go on getting promotion to a very high level, and still be a specialist in your own practical technique . . . . In the Australian scientific civil service, rank is independent of position in the organization . . . . That is a very nice solution to the problem'.

They also appreciate the working atmosphere in an establishment that:

'isn't hierarchical, in that just because you become a PSO doesn't immediately put you over evrybody else who is of lower grade . . . . The individual merit promotion idea helps that, in that it means you don't have to accumulate this empire to be promoted, and because people are a different grade it does not mean to say they have to look up or down to each other'.

On the other hand many are 'unhappy with the special merit situation' as at present administered. They complain, for example, about significant inequities: 'Why is it that [establishment X] has so many special merit SPSOs, and [this establishment] does not?'. Attention is drawn to undeserved appointments, or to 'the chap who has really done nothing since he was appointed'. Some of these complaints might be interpreted as typical rationalizations of lack of personal success. But there is also a strong feeling that 'all promotions on special merit for SPSO are purely on the research side, and do tend to be in rather specialist areas', and that 'if you are doing other sorts of work where you have to change regularly from one job to another, then . . . you don't have the possibility'.

What they are saying, in effect, is that 'individual merit' as a scientist is not to be measured solely by traditional academic criteria. The functions that many PSOs perform *as scientists* are much more diverse than they used to be. Individuals who carry out these functions in an outstanding manner should be recognized and given the appropriate personal standing within the organization. Some of them would, no doubt, make good line managers, but there are others whose talents and experience can be used to good effect in many other ways, as project leaders, external advisors, repositories of technical wisdom, trouble shooters, and so on. The 'senior scientist' − by whatever criteria he or she is selected − has an important organizational role that could be greatly extended, but as yet:

> 'We don't think how we should prepare scientists . . . to become senior scientists . . . . We don't actually tell them what we expect of them at that level.'

As Holdgate (1980) remarks concerning the whole system:

> 'The present machinery for effecting promotion to SPSO and above . . . unconsciously militates *against* the longer perspective of career development. By stressing the need for the individual to do the job being advertised as soon as possible, there is an inherent bias towards the individual with specialist knowledge relevant to the post in question. The high-flying good generalist scientist may have been at a disadvantage.'

Dual ladder career schemes such as IMP are designed primarily to counterbalance the predominance of managerial criteria for the recognition and motivation of outstanding achievement in bureaucratic R&D organizations. Such schemes are becoming more common, and although 'their introduction is a popular subject of discourse about the management of R&D' (Griffiths 1981), they are by no means without their difficulties and disadvantages (Thomason 1970; Wolff 1980; Griffiths 1981).

It may be, for example, that dual ladder schemes do not, after all, resolve 'the paradox that extrinsic rewards (pay, status, etc.) cannot be relied on to motivate achievement, but that when achievement occurs the extrinsic rewards should be consistent' (Pelz & Andrews 1976). As Bailyn (1981) points out, 'the dual ladder is an alternative progression to the managerial track, and as such represents a choice point for professionals at an *early* point in their careers'. Indeed (Hirsh 1981) there is reason to believe that the critical phase of differentiation between specialist and managerial careers is around 35, when many jobs are 'half and half', and is thus already determined before the formal forking of the ways. It is interesting that one of the objections to the IMP scheme was that:

> 'Most people . . . do their best research work . . . between about 25 and . . . 40. This is just the time when they do not get promoted on special merit . . . . If people are going to be singled out for special merit, they should be singled out when they are at their most productive.'

One can think of many objections to this proposal, but it would at least permit the application of much broader criteria for the majority of 'normal' promotions to PSO (§7.5).

But the real need (Bailyn 1981) is for a 'package of rewards' that 'comes at a *late* stage in a person's career, and represents something a technical professional can look forward to over a long period of service'. This might, for example, include 'the right to report directly to a higher level of management', as well as more conventional rewards of pay and public status. Bailyn's concluding remarks are so apposite to our whole enquiry that they deserve to be quoted at length:

'An organization that institutes such career rewards would soon have a group of senior professionals who were well integrated into the organization and well disposed toward it. They could form what I like to think of as an ''experience circle' – akin to quality control circles – a group of senior professionals with long experience in the company who could consider issues of work effectiveness and productivity from a wide range of perspectives. Membership in such an ''experience circle'' would represent an assignment of high relevance to the company. The existence of such a group would underline the distinctive contributions that can be made by older professionals, and would enhance their effectiveness and satisfaction.

Perhaps you feel that I am suggesting advantages for older professionals at the expense of younger and presumably more energetic and more up-to-date employees. I do believe that here we really have a ''trickle down'' effect. Part of the discomfort of younger employees is that they see only one route to success, a route that they well know is open only to a few. By providing for them models of satisfactory careers of many different kinds, we contribute, also, to their effectiveness and well-being.

In general, these studies point to the conclusion that variety, flexibility, and imaginative alternatives to standardized career patterns are not distractions to organizational efficiency, and may well be the most necessary ingredients in a productive future.'

7.7 External contacts

'Ideas move around inside people' (Ziman 1974). Scientists know the value of personal contacts outside the establishment where they are employed:

'we obviously have a very pleasant place to live near, we have roots here, [nevertheless] we have such good contact . . . throughout the UK in all sorts of different establishments, and thoughout the world, we come into contact with all sorts of people, and I don't think we are sheltered and getting into a rut in any way. I don't think there is much danger of that'.

Foreign visitors are encouraged:

'Some of the most welcome visitors are those who come on sabbatical leave from universities, who are probably quite well known in their field, and bring their special expertise and insights to the lab . . . they often bring . . . new thought to your own research problems . . . [they] come partly to write, and partly to do research, and to get involved in somebody else's problems. This is very good.'

A positive attitude towards such contacts is a subtle indicator of excellence in an R&D establishment. External relations with other scientists in other establishments can be regarded as 'ersatz mobility', guaranteeing its intellectual autonomy (Lemaine *et al.* 1972). Conversely, isolation is both a cause and a symptom of lack of quality:

'. . . we've had people who come along and say, "How is it visitors never come and see me?"
"You never . . . get visitors! . . . . Well, visitors go where they want to. You have to draw your own conclusion from that. If you want visitors, then you have to think about why visitors aren't coming to see you!".'

A feeling of inferiority may even induce a research group to avoid 'dangerous' contacts with more competent groups (Lemaine *et al.* 1972).

External contacts are indispensable for the intellectual vitality of every scientist:
'How does one learn to evolve during a period of, say, nine or ten years' research? . . . . Visiting other establishments over a period abroad is really one very important way of doing [it]. Mobility is an integral part of changing one's research methods and approaches, over the years.'

In some ways, scientists nowadays are not so 'compartmentalized' as they used to be:

'. . . instead of staying in the lab, as we would have done in the past, [we are] getting out and really mixing the soup . . . getting in at the various establishments . . . . Going into industry as industry has never come to us before . . .'.

Commission research widens the range of contacts:

'. . . the peer group also includes customers. Just as a doctor can have the approbation of his fellow professionals, so he can have the approbation of his patients. When one deals with a lot of people, a lot of situations, then the feeling that [one has] the confidence of the customer, of being taken into his counsels . . . talk about his intimate difficult problems, and then the feeling of gratitude when one has made a contribution — I think that is a very good peer group to be in . . .'.

Industrial researchers say that they 'never had any problems at all in getting into universities' to establish links. In a large multinational company, researchers may also be involved in a 'very close-knit network' of world-wide interchanges of technical knowledge. In some R&D establishments, the nature of the research may necessitate very extensive overseas contacts:

'The people within our unit probably do more overseas travelling, and more liaison with groups overseas than any other group.'

But most scientists establish and maintain personal contact with their scientific peers by attending national and international scientific meetings. These range in scope and scale from highly specialized weekend 'workshops' for a couple of dozen carefully selected participants to monster congresses where thousands of members of a whole discipline or subdiscipline congregate for a week from all over the world. They take place so frequently, and cover every subject from so many different aspects that it is impossible to attend all the ones in which one might be interested.

An active research scientist cannot hope to keep in touch with the current ideas and significant people in a specialty unless he or she is able to attend a selection of such meetings, at home and abroad, each year. Unfortunately, this would cost a fair amount of money − several per cent of their personal pay − if they had to pay from their own pocket, and the current climate of financial stringency has withered institutional travel funds:

'The money available for overseas travel has become very seriously limited. In fact, this year, the amount of money for internal travel [within the UK] has been quite slim, but the last time I went to an international conference was [three years ago]. Money has got steadily worse since then.'

Scientists themselves sometimes blame 'higher levels', where there is 'a general feeling . . . that unless you are actually sitting at your own desk it is not work':

'The criterion seems to be that you are writing a paper which is boosting the [organization's] morale in the outside world . . . . If you are involved in a project in which they have invested quite a lot of money, and you have results on, then I think you have a fair chance of going. If you just want to go abroad for learning, they would say ''Well, there's a library upstairs'' . . . .'

' . . . There is an attitude that it's not the cost, but it's the fact that you are away from work for a week . . . . A week at a conference might be much more fruitful than a week working here.'

But even when it is generally agreed by management that 'it's a good idea to travel and meet contemporaries overseas', getting funds has become 'a tortuous business', which fails to meet the real demand.

' . . . We've got so many people going [to a forthcoming conference abroad] by actually driving ourselves in a mini bus. We only had a budget for about four, and instead of going by public transport and paying the full rate, we pooled this money together . . . hired a minibus, and got seven people going.'

Attendance at conferences is not only an efficient way of extending ones scientific contacts:

'I had an excuse to go to the States . . . . I went to a conference on [a subject I was interested in] . . . so I managed to find the people working in the area there, and get together with them, and pick up threads.'

It can also provide 'the opportunity to visit one or other laboratory' at minimal expense. Rather than putting in a 'request to . . . visit such and such a lab to speak to so and so who is actually doing the work . . . the much easier way is to find a conference overseas that is applicable to your work as well, and tie on visits to laboratories, and visits to other staff as a part of that'. Such visits are highly valued, even when quite short:

'I had a fortnight in the States, to visit two laboratories doing similar work to ours, and that was the type of trip that . . . we would like to do . . . to go over and speak to people who are doing similar work, and see how they are going on.'

What is often needed, however, to break away from an old specialty into a new one, is to spend many months — even a year or so — in another intellectual environment. In the past, many academics were able to do this as of right, by taking *sabbatical leave* from their university duties, and going to some foreign university to do research. Not surprisingly, shortage of funds has severely curtailed this privilege. It is almost always necessary to 'cover your salary' from some other source, and it is seldom possible to 'get some assistance . . . towards getting the family away'.

In principle:

'The organization should be that people should be offered a sabbatical, and if they wish to take it up then let them get on with it, and if they don't, then they lose that opportunity.'

In practice, however:

'the people who have gone on sabbaticals, and asked for it have tended to be those who wanted to go and dug it out for themselves'.

One obvious difficulty is that:

' . . . when you are a little older, and you have family commitments and all this sort of thing, it's not so easy just to uproot for a year or two . . . '.

As a consequence, in many R&D organizations, scientists in mid-career are not encouraged or helped to go away for long periods, however beneficial this might be to their work or their careers:

'We would have to find funding from somewhere else — the host establishment — and one would then apply for special leave without pay, with the disruptive influence . . . on the work within your small group. People who tend to go off tend to be [below PSO] level, where there is much less disruptive [effect] when they go. There are at least three or four people I can think of who have gone abroad for a year or two on this sort of basis, to work in a university or other establishment. It tends not to happen at our level . . . '.

' . . . I think you would have difficulty in convincing the Establishment that
   you were expecting them to pay you for useful development of your career
   within the [organization].'
As we shall see in the next chapter, much more positive and generous attitude
towards long visits abroad at later stages in their careers would help many
scientists into and through major episodes of change.

8
Managing a change

8.1 Initiating a move

However well their previous career may have prepared them for it (Chapter 7), a scientist making a substantial change of research specialty or organizational function is bound to find the transition challenging and traumatic. What can be done by 'the management' to provide guidance and support through this period (§7.1)?

In many cases, the change arises from a managerial decision that is not related to the career situation of the individuals affected by it. Management then has a clear responsibility to help each person through the transition. Consideration has to be given to the future of research scientists displaced by a major policy change or administrative reorganization. The opening up or closing down of an extensive, urgent, R&D project (§2.3) usually generates a large number of abrupt moves which have to be orchestrated from above:

'Some of them were voluntary, but a lot of them were picked out and were moved − not in an inhuman or unpleasant way, but . . . we had problems of morale. They felt a bit displaced, these people, and some of them were rather resistant to it. So it was a combination of people who volunteered for it [and] other people [who] were just picked out and told.'

Such situations are now much more frequent than they were a decade ago (§3.6). Given time and tactful management they can provide an effective mechanism for individual and institutional development:

'I think it is fair to say that [some of the staff of the establishment] were drafted in [to this team]. It is also true that the ones who were drafted in did not kick against being drafted. They were quite happy to go along with it. The challenge − the fresh type of work. It so happens that the ones who were associated with it had been in some similar work in the past, so they were able to adapt more easily.

. . . It was one of those situations where it evolved. The initial stage was some various discussions, recognizing that there was an area of research into which we should develop, and who was interested . . . .

. . . We recognized *then* that it was going to be a particularly technical subject . . . . That pushed it very much toward [ ]'s area, and then . . . one or two others showed interest, and there were opportunities for staff to be drafted in. Some of the people were fairly obvious . . . . When it was mentioned to [–] he was interested quite quickly.

. . . [–] was not drafted in. He asked if he could become involved in the work because of his interest in [an aspect of the work] and his past experience.

. . . We had great difficulty in filling [a particular technical post] . . . . In the end one of our [staff in a neighbouring specialty] volunteered for it. We had intended asking him if he would, but he did sense that there was this interest, and he volunteered for this challenge, and he moved [his specialty].'

Managerial action may also be involved, from the start, in purely individual career transitions. It is true that a scientist may recognize the need for change without prompting (§5.6), and personally initiate the transition process. Self-induced moves of this kind are not at all uncommon in academic scientific careers (§1.5, £1.6), although they are not always welcomed or facilitated by the powers that be. But most people find it very difficult to take such decisions without some external stimulus – a stimulus that friends and colleagues are often reluctant to provide voluntarily for reasons of courtesy.

A research manager who is in a position to take the initiative in sparking off a major change in another person's scientific career clearly bears a good deal of responsibility. The first indications of 'staleness' (§5.6) can often be detected by others before they are evident to the victim, but any suggestion of making a change is liable to be interpreted as a slur on ones competence (§5.5). On the other hand by the time the symptoms have become obvious – e.g. '3 black marks on annual reports' – it is usually too late to do anything about it.

Unfortunately, line managers in R&D organizations often resist their staff being moved, and are not always very good at assessing the aptitudes of scientists for different types of jobs. Yet they are usually the only people who have a thorough grasp of the nature of the highly expert work that may be in question:

' . . . You have to have a a professional view, as it were, of the [person's] deficiencies. The problem all the time is that the level at which you need to have a clear view of whether [they] are making hay or not is at the professional level. When you decide to do something about it . . . then you've got to look around and think of a way of doing it. But in the first instance, the notion that somebody has gone stale, or that they are not addressing the right issues, or whatever it is – that's a very profound professional judgement.'

Such decisions may be difficult, and their immediate consequences embarrassing, but the long-term consequences of shirking them can be much more unhappy.

8.2   What are the options?

It is one thing to know, or be told, that one ought to make a change: it is quite a different matter to decide where to move. Most of the immediate, practical arguments are for staying put (§5.2 − 5.4)! The nature of the decision will depend on how strongly one is being pressed: is the change 'expected', or 'required', or will it be 'enforced' by management? In any case, what options are really open? A managerial post may be desired, but there may be little prospect of promotion (§6.1). There may few openings for a sideways move into an administrative position (§6.3), and technical service work (§6.4) may not seem very attractive. A change of research field is the most likely possibility.

Ideally, the research scientist is a self-winding, autonomous individual, whose choice of a new problem area ought to be entirely personal (§3.4). In practice, this choice is limited by a variety of institutional factors − the availability of apparatus and collaborators, the research programme of the establishment, etc. (Ziman 1981). Even when the move is not 'enforced', managers need to act positively by indicating a realistic range of alternatives:

' . . . You have to offer them possibilities. It's really a question of auto-suggestion, isn't it? Suggest to them that they suggest to you that that's something they ought to do. To do more than that, you've got to suggest that their continuation of work that they were doing for 15−20 years is unsatisfactory, and that won't do any more. That's the first point. Then you've got to offer, at the same time, the possibility: but at the same time, don't let them think that their talents are of no value in the future.'

Whether advisory or mandatory, any such proposal obviously rests upon a careful and thoughtful assessment of what the person in question can really do. Intangible factors must be taken into account (§4.8):

' . . . You run a mix of people . . . . Certain situations . . . certain ways in jobs, suit people, and you really have to match the person and the job, to get the maximum out of people.'

But such factors can only be assessed very subjectively:

'[When I moved to this new job] I found that the management were extremely helpful . . . . They suggested it to me for the purpose of furthering my career, because they felt that the emphasis in my previous place was far too engineering for my temperament, and that I ought to be on a project which . . . enabled me to be theoretical and not to have to worry all the time about the immediate [problems].'

Too much weight should not be given to the popular notion that a person's job needs to be matched closely to their 'personality', as assessed by systematic psychological tests. A study of mid-career changes (Thomas & Robbins 1979) found that people did not move into careers more congruent with their personality type, and if they did they were not more satisfied with their new careers. The main result was that the change itself was 'very rewarding', whether it was to a more or a less congruent type. This caveat is particularly important in relation to factors such as 'versatility' or 'adaptability', which, if they are indeed innate traits of personality (§4.1), may simply never have been evoked by past experience.

On the contrary, for a highly skilled research specialist, technical considerations must come first. The basic question is whether a person *knows* enough to take up a new specialty or a new organizational function. This is the most elementary concern of R&D management:

' . . . I think we are fairly well documented. I mean, there is a system which is for ever pulling us up and looking at the roots, and I think it is fairly well known where our strengths and weaknesses lie, and I suspect that if someone asked you to move to a job which was outside your strength . . . then people would be extremely unhappy about it . . . . I have never been asked to move to a job which I haven't been able to cope with . . . I don't really think that is me: I think it's the people have been looking, and saying . . . ''he can do this sort of thing'' . . . .

. . . When people are moved, positively moved, it's generally successful, and it's generally the right people, who are able to cope.'

Nevertheless there is a danger of underestimating the range of skills that a scientist may have acquired in the course of his or her research career (§4.4). This is the key to a successful move:

' . . . As you move from one field to another, you have to take at least one skilll with you. In other words, you can't be a fish totally out of water . . . moving into [a new area] I brought my [−] technology with me, and [was] able to apply that to this other area, but at the same time then pick up skills relevant to the new area . . . . The versatility of people depends . . . on whether at least one of their skills is transferrable between the jobs.'

True managerial expertise is to recognize the existence of a transferrable skill amongst the many that a scientist may have:

' . . . I can take some of the blame . . . for steering X towards that [topic], for I was aware of some other areas of work [at another establishment] which required [a technique] of a certain class, which we had been interested in and previously was his field . . . . So I steered X towards that . . . [He] did a literature search, and . . . found his own problem to a great degree, and then became independent.'

It is not enough for the manager to know that someone has some competence in a rather general skill such as computing: it is necessary to be able to point to their 'background in the study of [] systems' and to say that they are 'just the right chap to work on [an analogous phenomenon in another system]'. This may seem a purely practical consideration, but there may be a deeper psychic commitment (§3.4) which has to be sensed and respected:

' . . . You need to have some thread that you maintain when you do change fields, which is something which I also recognized when I moved about a year ago. It is not something which is explicitly examined in any way by the management; it's up to you, the individual, to make sure that that is preserved.'

8.3 Retraining

It is quite normal, nowadays, for skilled workers to go through a formal course of retraining in order to take on a new type of job. Is this what scientists should do when they change their fields of research? After all, researchers are often recruited directly from universities (§3.3, §7.2) where they have just been through up-to-date MSc and PhD courses in the relevant specialties. A scientist who is changing field would normally have a first degree in an appropriate basic discipline (§4.3), and would thus be qualified to enter such a course and learn a new specialty systematically before beginning research on it.

From the limited viewpoint of a project manager, putting a lot of effort into formally retraining a scientist to fill a particular slot may seem too much of a gamble:

'First of all, you have got to find somebody who is willing to do it; then your have got to find out whether he is capable of doing it; and then, at the end of the training period, you have got to find out how good he is. Now he may be willing, he may be interested — and he may be a dummy. That's the fact of life you have to face. If you take in the staff from outside, if you search around, you interview, and you employ somebody to do the job from outside, you get somebody . . . with a proven record of known ability. If you take a member from within your organization, and try to retrain him in something too different, the outcome is in doubt. Maybe useless.'

On the other hand, at a higher level of management, different considerations may apply:

'We did not have any proper commercial experience . . . at the time. There was a decision made by the [establishment] management. Should it be to bring in professional people from outside to lead this marketing. They would then have people who did not understand [the establishment] and did not understand science. Or, alternatively, they could try and retrain scientists

from scratch . . . they chose the latter course, because they thought people
would be better from a motivational point of view. I think it worked.'
Formal retraining is certainly successful in some cases. A scientist with a PhD
recruited into a technical job in a totally new field found that it was 'an art . . . not
a science':

> 'and for nearly six months I was very disillusioned, because I thought that
> I wasn't getting anywhere and I wasn't learning anything . . . . I was trying
> to do a lot of reading and catch up with all these terms . . . . The only way
> I thought of doing it was getting a formal qualification, and so I convinced
> my section manager to send me on an MSc at []. [My employers] were very
> kind, and supplied me with money for nine months to go away and do that,
> and I came back and I have just nearly completed all the exams . . . '.

But there are legitimate objections to the opinion that 'we do not do nearly enough
of the formal retraining in new subjects'. For example, existing courses are
seldom well matched to immediate requirements. Thus, when a completely new
research programme was to be launched and a major project was 'grinding to a
halt'

> ' . . . there were certain staff that were obviously in need of redeployment,
> and these two things became enmeshed, and six people were identified at [the
> establishment] to undertake a certain amount of training at [a major
> university] . . . . [This was] a six weeks' course . . . we had the first
> fortnight with the general MSc course on []; the rest of the time was spent
> on a course specifically designed for our purposes'.

In many cases:

> ' . . . there is going to be a limited number of places that we could have gone
> to . . . .
>
> We could have got the basic retraining, I suspect . . . but it would not have
> helped us very much. I am sure we could get that from reading, anyway. We
> would have got the retraining if we were prepared to go to a very large
> number of different establishments. *We* could have gone to one . . . *you*
> could have gone somewhere else'.

In these circumstances, it may even be advisable to 'bring in consultants
and . . . run courses' within the establishment to 'put people through a very
onerous brainwashing, to train them and give them some professional
background'.

The fact is, however, that experienced research scientists seldom learn new
skills by attending advanced courses of instruction (§4.5). They may go to an
elementary, introductory course in a novel subject just to 'get the hang of it', but
they acquire most of their real expertise by 'learning on the job'. They are habitual
self-educators, accustomed to teaching themselves directly from books and

articles or by informal conversation with expert practitioners (§4.6). They already have tacit understanding of scientific work (§4.7) which makes them impatient of courses designed to train beginners. A course of formal retraining may sometimes be what is needed to make the transition to a new specialty, but generous encouragement and opportunity for self-generated personal development may be just as helpful.

8.4   Facilitating the transition

The process of transition from one specialty to another is often very confusing and disheartening. There is no simple managerial recipe for facilitating this process except sympathy and support. This applies particularly to finding a way into the personal 'network' of a specialty (§5.3):

' . . . If you become established in a particular area, you are competent and
confident in that area, and you can talk to anyone in your particular field of
expertise in comfort. But [if] you leap into a new area . . . you may have
a certain amount of self-confidence, but it's not really justified. You don't
know much about it, and I find difficulty getting myself to the point where
I could . . . discuss ideas as an equal.'

Leadership from 'other scientists who had some experience in the work' can obviously be 'very helpful in bringing [one] more into the game'. Exploiting previous contacts and developing interactions with scientists in other establishments 'who were already involved in this area' can make the transition much easier in a field where 'there isn't a vast amount of well-documented work'. At this stage of the process, it may not be a bad thing to have a certain amount of work from previous projects to finish off (§5.4), simply to keep one fully occupied whilst new ideas are slowly being assimilated.

Otherwise, it is mostly a matter of feeling ones way, step by step:

' . . . There are meetings I would go to . . . where experimental things
were discussed, and I would pick a name out of that, and then follow that up
at another meeting . . . . You slowly develop a list of names of people doing
interesting things. From the literature you also come across it, and if there
are organizations with which you can establish good links, then visits there
can be very valuable . . . . Two or three days [at a major centre in the United
States] is so valuable when you are in the formative stage, and they've got
experiments there, and they are doing things, and you can say why, and they
tell you why.'

A mature scientist who already 'knows how to do research' (§4.7) will not be discouraged by the fact that:

'. . . half the time is developing thoughts, because it's only slowly you can pick up ideas from other people that they have published in the literature, as you go to visit various laboratories'.

Managers may always have 'a feeling you send people away, and then they never come back again', but longer visits (§7.7) can be put to good effect:

'I went to [another research centre] specifically to do a small amount of research, but the idea was to spend a lot of time talking to the [people] there, and reading, and attending lectures that they were giving. But really it was just to work with them, and to sort of accrue the information by actually having to put it into practice. It's probably a good way . . . having the people there whom you can ask the questions of at the time that you need to ask them.'

Strangely enough, the most successful transitions have been into quite novel fields of research. A newly developing specialty is not only attractive in itself (§5.7): it can be psychologically less daunting:

'There may be a certain diffidence in people . . . that we must get out. If they go into an "establishment", they may not be wishing to make any contribution, because when they first start they are not really capable of making much of a contribution, and they tend to be a little worried about putting any ideas forward, in case it's howled down or something . . . . In our most recent change . . . we've all got into something which has been just beginning, so . . . we have been able to make our own mistakes without having an expert on the top pointing out how damn stupid we were.'

A self-winding scientist, accustomed to considerable autonomy in the practice of research (§3.4) can even get some satisfaction from a situation where:

'. . . in common with everyone else, the new techniques I had to learn, new disciplines, were learnt by talking to other people who had been in the field a week or two longer than I had . . . . And reading? . . . . There were no textbooks on the subject, there were conference papers. It was not the sort of work which one could expect to learn the techniques by going for any formal education of any sort'.

8.5　Getting re-established

Both for the person making the change, and for his or her managers, it is absolutely essential to have a clear notion of the time scale of the process. In a profession such as the administrative civil service or industrial management it may be appropriate that 'a guy is in a [new] job at nine o'clock, and line management would like to see him able to do the job by ten'. But where there are technical judgements to be made 'it takes up to six months to get used to the new job: in that time you are really only doing half a job or a quarter of a job'. To become really expert, 'you've got to see every problem at least once' which means 'you're talking about three or four years'.

In a narrow research specialty, it would seem quite obvious (§4.5) that 'you can't create an instant expert: no way!'. Looking at a change of specialty as if one were starting from scratch, one would expect it to be a very long process:

'A person needs three years' university, three years' postgraduate, and, say, then three years' post-doctoral experience — nine years' training. You can't become [a specialist in X] — you can't expect somebody to go and pick that up in a year. You can't. That's impossible.'

But an experienced research scientist is not starting again at the level of a student (§4.7). The transition process often turns out to be much shorter and less stressful than it appears in prospect (§5.6):

'. . . having had a basic physics training . . . one is equipped to do all sorts of things in the . . . physical sciences area . . . . For six months you've got to read, you've got to think, you've got to consider what is going on, you've got to educate yourself. But it was by no means a painful process . . . .

[The new project] was . . . fairly straightforward. A certain amount of literature existed on it, but inevitably one has to do a certain amount oneself. But it wasn't a sudden transition. It would not be like suddenly . . . saying "Well, look, tomorrow I want you to go to [the local] Hospital. You will be the major abdominal surgeon in [that] Hospital. Take your tools with you if you can". It's not that sort of change . . . .

. . . it's an exponential curve . . . . After six months to a year, one is starting to become quite . . . . It's a very slow . . . you don't quite realize . . . the rate at which you are accumulating information . . . '.

Scientific specialties are not as impenetrable as they seem to the outsider (§5.7):

'. . . . I have always been amazed by what people do . . . . I have seen a lot of cases where people have moved into a new area, and they have just staked a claim, maybe in a a university department — "I am going to work on this" — and it's remarkable . . . how quickly they establish themselves as one of the leading people in the area. Getting invited to talk at meetings — at times a fresh face scores . . . — and within a couple of years, usually — they've made a few bloomers, yes, but they fairly quickly learn the ropes — and in no time at all, they know as much as the . . . oldest and most revered member of the group, almost.'

Even the mature scientist who already knows 'how to do research' (§4.7) may get bogged down at first:

'. . . the first thing was to ask what one earth you should buy or do . . . . If you are not in the field yourself, it is very difficult to know who to ask. You have to start going to universities, and if you know very little at all, it is very difficult to know which universities to go to, what people to ask . . . . You can't really expect to spend a great deal of their time explaining to you all about [the new subject] . . . . So you have to start reading books about it, and you don't just learn thing [so much by] reading books as from a practical application, and so it has taken a great deal of time to learn about the sort of things that we need to do . . . '.

But such a period of frustration is not necessarily wasted:

' . . . I started from the literature, to see what other folk had done, and buying a whole lot of gear, and trying to put the Labs. together . . . . Eventually, at the end of the year, I had a rough idea of what I was going to do basically, and what other people were doing. The next year I got to make contact with a lot of people, and realized that the majority of the work which had previously been writted up in this field was a load of rubbish. It's absolutely appalling . . . . So that I started off based on the literature I looked at, and was intending to do the same sort of stuff . . . I am going to have more thinking to do about the basic problem, possibly, than I should have done when I first started off. I just did not know any better . . . '.

A time scale of 'about two years . . . to actually get to grips with the sensible approach, let alone actually getting out a coherent result' thus seems typical:

' . . . we are beginning to get places now, talking of being 12 months after we are actually allowed to spend money − about 18 months from when we first got involved in this at all. So it's something like a two-year timescale to get yourself established'.

In a discipline where each experiment takes some time, the hiatus in the flow of publications (§5.3) may last for four or five years, because 'one has to do the work spanning several years', even though one can begin to 'make progress and understand the problems and . . . get information after about three years'. But the researcher's own criterion for a successful transition to a new field is that they 'could probably hold [their] own with anybody in [that field]':

'It's a . . . gradual process: one could not even say, now, if one is right up-to-date. We did a six-week course. Really that was just to overcome the intellectual shock . . . . You learn the jargon at that time. We then went through a year [replicating a previous investigation] . . . . Having trod the footsteps of [others] we were able to see the [assumptions they were making], and so at the end of the year we felt reasonably competent in flinging ourselves into, perhaps, a new study, asking the right sort of questions . . . . After a year and a half, two years . . . we felt . . . we were on a par with many [R&D organizations in this field]. Some [R&D organizations] are much better than others, and . . . we have perhaps reached a medium range of standard, and from there on it's just a question of learning further and further tricks of the trade.'

As the above quotations suggest, a researcher making a mid-career change can get strong support from membership of an active team (§7.4) responding to the challenge of a new type of problem (§5.7). A move into a field where there is already local expertise may also be relatively easy, although:

' . . . When you make these changes, you go into a sea of experts in the new discipline, so when you make a contribution you are supported by them. The real crunch comes when you think you are in that discipline, and you make the decision to move somewhere else where there may not be those people, and then you can't function in that discipline without that massive knowledge base.'

But these circumstances are incidental to a basic process of psychological adaptation which must be allowed time to develop. Insistence on very rapid change is not consistent with the technical autonomy which scientists value so highly (§4.8). Managements may feel that this time is long by comparison with the 'urgency' of some R&D projects (§2.3), but if the transition is successful it need not be a disastrous break in the normal course of either a reputational or an organizational career (§3.3). The basic message comes from a scientist who had recently 'changed quite dramatically' and said that 'it took quite a long time . . . to adjust': in fact, it was only a matter of 'two years, really, before I really settled down, and was happy and content with what I am doing now'.

A manager who fully understands these things, and can provide the right sort of support through this period of uncertainty can sometimes do wonders:

' . . . People get into a rut. People also persuade themselves that they can't do certain things, or might not be able to do so. But . . . if I think of the team with whom I work right now, a number of them were presented to me as — well — "Look, old so and so. You urgently need somebody to turn a particular handle. Old so and so, he is getting near to retirement now," or "he is hopeless", or "he is absolutely the last word". "You can have him for nothing. Look, he won't be charged against your project number. You just look after him . . . . He is not a bad chap, but, you know, he is no good." . . . I actually tell the people who were presented to me in that way what useless louts they were presented to me as, but . . . they have come up absolutely trumps . . . . I was able to motivate those people into doing something, and they like it . . . .

. . . one of them was a refugee from [a discontinued project] . . . . He had been put into the wrong sort of environment with somebody who did not really understand what was going on . . . . He is going to retire next year . . . and we are dead worried about his departure.'

9

Organizational policies and personal roles

9.1 Management dilemmas

A career change is a highly personal event which has to be handled primarily at the lowest level of local management. But the conditions under which such changes take place are set at higher levels in the organization. This applies obviously to general policies for recruitment (§7.2), diversification of experience (§7.3), promotion (§7.5) and job definition (§7.6), as well as to systematic job appraisal and career guidance (§7.1) and to many practical administrative details such as the transferrability of pensions (§5.8), allowances for domestic upheaval (§5.8), opportunities for travel (§7.7), etc.

The most important issue for top management is the balance between 'redeployment' and 'redundancy'. How far is it desirable, or possible, to deal with personal and institutional change by moving people around within the organization, instead of terminating the employment of a certain number of individuals? The present study is concerned solely with the means of putting into practice a systematic policy of redeployment, and does not attempt to explore the mechanisms and consequences of making people 'redundant' in mid-career. As already indicated (§3.6), I share the attitude expressed in one of the interviews:

'My starting point is a humanistic one, which is that organizations like the research councils have got no right to simply work on the assumption that they can export people at 45 to somewhere or another, and forget about them. When I talk about careers and professions, that means that people have a reasonable right, entering such a profession, to a relatively open path through to retirement, in productive work of one sort or another.'

The senior managers of most R&D organizations would subscribe wholeheartedly to this basic principle. Indeed, until recently, this was the general practice, even in the industrial sector, where there is a longer tradition of positive career management (§7.1), including the power to dismiss people for unsatisfactory work. For example, the European Industrial Research Management Association reported very low staff turnover rates in industrial R&D, and practically no dismissals. 'Any idea that industry in its research establishments (in Europe) operates a "Hire & Fire" Policy is quite wrong' (EIRMA 1970).

But for the last five years or so, many R&D organizations have not found it possible to follow this policy, and have been forced willy-nilly to dismiss some of their scientific and technical staff. In 1981, when we were conducting interviews at several major industrial firms, this was still considered to be a novel situation. But in one such firm it was thought that about 10 per cent of the R&D staff would have to go — mainly managers in their 50s who had not successfully transferred to other parts of the organization — with little possibility of finding further employment in another firm (cf. §7.2)

The more rigorous administrative and financial climate in which R&D organizations must now operate (§2.6) has shifted the balance towards 'redundancy' at two levels. The direct effect has been the enforced run-down, closure, or transfer of title (e.g. 'privatization') of a number of R&D establishments in the research council and government sectors, at a greater rate than could be accommodated by the redeployment of their staffs, whether voluntary or involuntary. Universities and polytechnics are under similar pressures to make drastic economies in their budgets for teaching and research.

Research is highly labour-intensive: abrupt and drastic budget cuts or sudden and sweeping changes of research programme can only be met by dismissing a number of employees at correspondingly short notice. Such episodes are so disruptive and damaging to research organizations and to research careers that they can only be justified in a situation of political or economic crisis. This is clearly an issue of high policy. It would go beyond the present study to comment on the current situation in the United Kingdom in this respect.

But this more rigorous economic climate has indirect effects that percolate down to lower levels in the R&D system. In particular, there is a continual demand for greater 'efficiency', both in the choice of problems for research and in the way that they are tackled. The whole business of research ought (so it is said) to be better managed. Institutions and individuals ought to be more flexible and adaptable, in order to undertake more 'relevant' and 'urgent' projects (§2.3). Resources ought not be wasted, through personal incompetence or idleness, or administrative ineptness — and so on.

Redundancy, or the threat of redundancy, is thus regarded in some circles as an essential managerial instrument — as the means for 'cutting out the dead wood', as the stock metaphor so charmingly puts it. This hard-headed point of view has its attractions. People always seem to be doing the wrong sort of work, or falling far below their supposed capabilities: it would be so much easier if they could be dispensed with or replaced. The difficulties of killing off a faltering project or closing down a second-rate unit are compounded by the necessity of finding suitable work for all the people who will have to be displaced.

All the time (§1.2, §3.6):

'[The subject] is changing, and we are at the point of change
which . . . some of us are going to be able to cope with, and some of us
aren't. This is a problem for old institutes [nowadays] that if the character
of the subject starts to change you can't bring new people in.'

In the past, the steady expansion of the whole research enterprise automatically
drew in cohorts of young people trained in new specialties. Now that this expansion has stopped, there is an 'accumulation of older people' (§3.5). Indeed, in any
organization, if redeployment policies are too effective, they create problems of
lack of recruitment that can seriously affect the age and balance of work force (Hutt
1981).

Senior R&D managers are very conscious of the institutional consequences of
immobility (§5.6). It is not so much that individual scientists 'inevitably burn out
by their mid-40s, but that they cannot be kept alight in a stagnant atmosphere. A
fall in staff turnover. from, say, 7per cent to 3per cent per annum (Decker & van
Atta 1973) presents managers with problems which are fully appreciated by all
working scientists:

'No management system can help you if you have no new blood coming in,
and no people going out. You are stuck. You're finished.'

These managerial considerations are cogent, but they do not obviously outweigh
the basic arguments for 'tenure' (§3.5). In all R&D, there is a creative tension
between the challenge of insecurity and the self-assurance of tenure (Pelz &
Andrews 1976). Scientists naturally value security and stability of employment
as a personal benefit, but that does not invalidate their claim that the intrinsic
quality of their work also benefits from these conditions.

It is certainly difficult to believe that the current practice of employing
researchers on a succession of short contracts (§3.3) is producing better, more
dedicated scientists:

'They always were accumulated by the way . . . . They were accumulated
on short-term contracts, because nobody would take the decision to sack
them when they could have. So you've got unbelievable people of 55, on their
12th three-year contract . . . . Now, of course, you change to a situation
where, in theory, you could get rid of people . . . and the question is, have
you got the same people? The argument has to be to justify that you've not:
that you've actually improved the quality of the people you are getting,
as . . . you're setting up these obstacles which they have to surmount in
order to get tenure . . . .'

The time to appraise people seriously for performance and promise (§7.3), and
to get rid of them if they are not up to standard, is in their 20s and early 30s, when
they could be retrained for another profession, not in their 40s and 50s when their

opportunities for alternative employment are much more restricted. The present fierce competition for regular research posts would justify a longer and more rigorous period of probation before confirming tenure. At the same time, the criteria for promotion might be made much stiffer, inside the 'tenured' group, so as to create a class of 'permanent juniors' to carry out more technical functions (§6.4) or to undertake commissioned research projects of a more routine character (TNO 1980).

Another general policy that can loosen up the research profession is to make generous provision for early voluntary retirement. It is generally reckoned that this becomes too costly for staff below the age of 55, and there is always the risk that the opportunity will be seized precisely by those able people who can easily find another job and who can least be spared. But scientists who have 'gone stale' (§5.6) or whose work is otherwise very weak are seldom unaware of their situation, and may even seek early retirement if the terms are known to be reasonably favourable. Every move of this kind makes a vacancy for the recruitment of new blood.

It is important to distinguish between 'the fact that people want to . . . economize people's jobs' and the argument that institutional efficiency and flexibility would improve markedly by making one lot of people redundant and hiring new ones in their place. The overall benefits of such a policy should not be exaggerated:

'It's not just . . . the scientific domain that turns out what you call scientific failures. I can think of lots of people in industry that strike out on this route much earlier. They get a position, they have a reasonable competence of doing something: if they couldn't do anything, they'd probably get shifted. There are certain advantages to keeping them there. They know a lot about it, they have some experience, and typically they are kept on to a certain sort of level. Now, if you are going to ship all these people out of all these different domains, where are you going to put them? . . . . It's true not only for the scientific community; it's true for the business community itself . . . because the people I've been thinking about earlier – there isn't anywhere to put them if you apply the same standards as we are applying, because in some sense you've got them operating in their area of best competence. What you've got to assume is that there is someone sitting outside, who, if he was transposed into that position in the lab, he'd be an all-singing, all-dancing researcher – and in fact no one has yet made that case.'

What this study shows is that many of the standard objections to the 'redeployment' of scientists in mid-career are not well founded. They have a wider range of skills (Chapter 4), and are much more willing to take up research in new fields (Chapter 5) than is usually reckoned. In many cases, it is the fault of their managers if they have been allowed to become too narrowly specialized (§5.5), or have not been adequately challenged with new tasks (§5.7).

The decline of personal commitment attributed to individuals is often due simply to organizational stagnation. Every R&D unit or establishment probably ought to be shaken up, restructured and redirected every ten years or so, but that does not mean that all its staff should be under notice of dismissal. There are better ways of getting them out of their ruts. Bringing older people together in a new research unit or a new interdisciplinary team (§7.4), for example, may be just as invigorating as an injection of 'new blood'.

This is not to minimize the managerial difficulties of putting redeployment policies into effect. The more rigorous economic climate has closed off some of the traditional outlets for scientists in need of a change of specialty, function or employer. It is no longer possible, as it was in the expansive 1960s, to move the weaker ones into the 'second ranks' of research. The units and establishments that now have to be closed down (§2.5, §5.6) are often much larger than they were in the past, and there are simply not the vacancies in other units, or in academia, to accommodate the people to be moved. In some establishments there are surpluses of overqualified people of high ability doing mundane research, and yet (§5.8):

> 'the opportunities for moving between different [establishments] within the
> same [R&D organization] are also very few and far between, largely because
> of the expense of moving people on any sort of permanent basis'.

More rigorous accountancy procedures get in the way of sound personnel policies:

> 'One difficulty that we do experience . . . is largely from the staff struc-
> ture . . . . You can have somebody who has achieved a reasonable degree
> of seniority, and expertise, and level of pay, and so on. If he is going to make
> a dramatic change, then it is unlikely for some time whether he can achieve
> the same level of competence . . . that justifies the salary that he is being
> paid. In this division . . . more than half are qualified scientists on com-
> mercial work, and therefore their cost have to be . . . justified'.

A structural problem that affects R&D organizations in both the research council and government sectors is their disconnection from other modes of expert technical employment — teaching, as in the universities (§6.5), production and commerce, as in industry (§6.3). There are simply not enough non-research posts for all the older scientists who could best be employed in them. This can be overcome to some extent by redefining the organizational role of the 'senior scientist' (§7.6), but it still remains a serious weakness in the modern practice of setting up full-time research establishments isolated from other social institutions.

This study was not designed specifically to look at the overall management of R&D establishments in a difficult period. Looking back, one can see that the constriction of resources for science in Britain (§2.5) was a slow, steady process, spread over a number of years. Nevertheless, it was evident from our group discussions that when it reached a critical level it caught many R&D establishments unprepared. More frequent and insistent redeployment of staff was only one of number of radical measures, such as increased diversification of programmes, greater urgency and relevance of projects, and a higher proportion of commissioned research, that were abruptly required of British R&D organizations to make them more flexible and responsive to outside pressures. My impression is, however, that some of them have weathered the storm very successfully, and that their scientific staffs have gained in self-confidence by the experience.

The managerial task they had to do was extremely difficult, because it takes time to transform the working atmosphere of an established institution. Scientific work develops an intrinsic rhythm, related to the time scale of its projects and programmes (§2.3). This time scale varies from institution to institution, depending on the urgency and extent of the work to be done. Scientists are able and willing as individuals to yield and adapt to sustained pressure for change on the time scale to which they are accustomed: it is quite a different matter to force a whole institution to make a sudden spurt and develop a more dynamic working atmosphere.

But some establishments were evidently not given much room to manoeuvre, and began to take crisis decisions on a day-to-day basis:

A. 'I threatened to resign only a year ago. That was because I was being switched from one division to another, just to fill a space . . . . It wasn't that I did not think one could do a useful job, but I very much resented the way I was being treated.'

B. 'I was also switched arbitrarily . . . called to the Deputy Director's office one morning at 11 o'clock, and suddenly I was doing different work the next day. But there was a rational reason for it — mainly that the laboratory was in a difficulty because somebody was retiring, and they had not had the foresight to do anything about it beforehand.'

A. 'I think that perhaps is the point as to how you treat people, and in particular scientists. There has to be a very rational, logical reason for what you are doing to them.'

Redeployment policies for research scientists are bound to run into difficulties if individuals are moved into new specialties without detailed consideration of their existing skills (§8.2) and without support during the year or so it will take them to make the transition (§8.5). Given these conditions, scientists turn out to be just as sensible and biddable as other people!

9.2   International comparisons

The issues we are concerned about are not unique to Britain. Laboratory life is very much the same the world over (§2.2), and scientists play similar social roles in all advanced industrial societies. The minute division of labour into technical specialties (§1.1) is characteristic of all scientific work, and shapes the careers of scientists in every country.

This is so obvious that it is sometimes assumed that all scientists, everywhere, follow the same career paths and face the same career problems. But this is not the case even within one country: it is clear, for example, that a scientist working for a large industrial firm is in a much better position to make a change of specialty or organizational function than a scientist with comparable research interests and capabilities in a small research council establishment. 'Versatility' and 'adaptability' may appear as individual traits, and vary widely from person to person (§5.1), but, as we have seen in Chapters 7 and 8, they are decisively influenced by social experience and are extremely sensitive to quite fine details of organizational practice.

Educational, administrative and cultural patterns within R&D organizations certainly differ greatly from country to country. A systematic analysis of these differences and of their effects on scientific careers would thus throw light on some of the particular problems encountered in Britain, just as some of the material in the present study should be of help to people dealing with similar problems in other countries.

The difficulty about such an international comparison is that it ought to be based upon comparable material − that is, on the personal experiences and opinions of working scientists in a variety of R&D establishments in each country. Unfortunately, the small amount of published material on scientific careers is only incidentally relevant to the present theme. Even a straightforward factual survey of practices and policies relating to scientific careers in a country such as France or Italy would be a major research undertaking if it were to get underneath the surface of statistical indicators and official regulations. The following remarks, therefore, are strictly impressionistic. They are not based upon objective evidence, and may well be quite mistaken on particular points. They are presented here, very briefly and unsystematically, simply to show the British situation in a world perspective and to indicate the range of issues that might come up in a proper analysis.

Generally speaking, the basic education of future scientists up to 'first degree' level is broader, and takes longer, in other countries than in Britain. Even a 'Master's' degree course at a typical American graduate school is designed as preparation for a profession rather than as an introduction to specialized research.

This postpones the decision to take up science as a career, which is peculiarly early in Britain (§4.3), and provides some diversification of experience at an impressionable age (§7.3). It is not clear, however, whether the broader general education thus provided has much direct effect on a scientist's versatility in mid-career (§4.3, §7.2).

The stage of 'apprenticeship' to the profession of research is nominally much more protracted in most other countries than it is in Britain (§4.3) In many European countries, it may take five to ten years to accumulate the material for a doctoral dissertation, by which time the candidate will already be established in a research appointment and fully committed to a research career. During that period, the candidate is forced to concentrate narrowly on the topic of his or her dissertation, and cannot afford to move into another field. The British practice of allowing only three or four years, from a bachelor's degree to a PhD, for formal training in research permits a further period of diverse postdoctoral experience which seems to make people more adaptable in later years (§7.3).

Recruitment and 'tenure' policies strongly influence attitudes towards personal mobility. In all countries, industrial R&D organizations normally recruit most of their scientists and engineers directly after graduation, and offer them more or less permanent employment. Industrial scientists thus have effective 'tenure' from an early stage in their careers, subject to the power of the management to move them from project to project, or even out of research altogether (§6.4). The extent to which industrial scientists actually change their specialties, or move from firm to firm, varies greatly from country to country. In this respect, American scientists are considered to be much more 'mobile' than their British counterparts, whilst Japanese scientists normally stay in the same firm all their working lives.

British, American and German industrial R&D establishments have always recruited a significant proportion of their staff directly out of academic research, typically after completion of a PhD. In France, until quite recently, this was almost unknown, so that there was practically no movement of people between the academic and industrial sectors of the R&D system. This institutional segmentation is even more extreme in countries that have built up their R&D systems on the Soviet model. Each specialized establishment of the National Academy of Sciences recruits its staff as individuals shortly after graduation, carries them through a long period of research training leading to the equivalent of a PhD, and provides them with secure employment until they retire. People do, of course, try to move to more prestigious establishments, or into more agreeable parts of the country, but this is seldom easy for practical reasons of accommodation, etc. (§5.8). The immobilism and segmentation of scientific work in these countries is often deplored officially, but seems to be an innate characteristic of the system by which it is organized.

Another feature of the Soviet system is the segregation of scientific research from employment in higher education, even in highly 'academic' subjects. Some senior scientists in research institutes do hold parallel university posts, but most university teachers have very limited facilities for research. As in some British research council establishments, the possibility of moving gracefully from active research into teaching (§6.5) is severely restricted.

By contrast, in the United States and in Holland (for example), it is not customary to set up separate full-time establishments for basic or 'strategic' scientific research (§2.3). Even more than in Britain, this is normally carried out in universities, either directly by, or under the supervision of, people who are also actively engaged in undergraduate or graduate teaching. Their careers are primarily 'reputational' (§3.3), although 'organizational' aspects may intrude more than is assumed in the traditional 'academic' conception of science.

In France and Germany, however, this type of research is carried out in relatively small establishments, mainly sited within universities but staffed by a corps of full-time, permanently tenured researchers, administratively separate from the university teachers as such. They too follow 'reputational' careers where specialized research publications are the main criteria for promotion (§5.3). Like the scientific staffs of the research councils in Britain, they have few openings for employment outside research, whether in management (§6.2), administration (§6.3), or other work suited to their technical capabilites (§6.4).

The part of the R&D system between the 'academic' and 'industrial' sectors is occupied in Britain mainly by governmental or 'quasi-non-governmental' establishments. The career patterns of scientists in the corresponding establishments in other countries must approximate to those of other state functionaries in those countries. It is an interesting question whether these form a distinct corps of 'scientific civil servants', as in Britain, or whether it is customary in some countries for some scientists to move across into other civil service functions in mid-career (§6.3). In any case, motives and opportunities for personal or institutional change undoubtedly vary markedly from country to country, in parallel with the diverse bureaucratic cultures in which these establishments are embedded.

The 'Rothschild' system by which government departments commission research from research council and governmental R&D establishments (§2.5), which now plays such an important part in the careers of many British scientists (§6.4), does not seem to be commonly used elsewhere. But in the United States the system of awarding research grants to individual researchers or small groups on the basis of competitive peer review is not confined to academia as it is in Britain, and funding agencies often provide openings for changes of specialty (§5.7) by inviting proposals for research in particular new fields. In France and

Germany, on the other hand, where mid-career academic and quasi-academic scientists do not have to rely so heavily on funds tied to specific research projects, they are not so strongly pressed to take up new subjects as these become scientifically fashionable or politically expedient.

This diversity of formal policies and practices is more than matched by a corresponding diversity of informal conventions and cultural norms. I can only hint at the significance of these by some obvious questions to which I do not know the answers. How is it, for example, that the professional trade unions are far more influential in the career decisions of scientists in France than they are in Britain? How are scientists in some American industrial firms encouraged to move so freely between research, administrative and managerial functions within the same establishment? How do Italian scientists protect their professional independence and personal careers against the excesses of a cumbersome bureaucracy? In what way do senior scientists in the Soviet Union use their considerable managerial powers to orient the careers of their subordinates towards change? Does the segmentation of German universities into small professorial institutes segregate scientists in narrow specialties throughout their careers? Is it easy to move between the various sectors of R&D in Holland and Denmark because these are small countries where 'everybody knows everybody'? Why do scientists in Spain drop their reputational careers in science when they are appointed to university posts?

Perhaps the questions to be asked internationally are even more general. Are the career difficulties attributed to excessive specialization common to scientists all over the world? If so, are they regarded as unavoidable features of research as a profession, or are they thought to stem from the particular way in which scientific work is organized and managed in each country? If the latter, is the problem thought to be as serious as it obviously is in Britain , and what steps, if any, are being taken to overcome it? There's a research project for somebody, if ever there was one!

9.3 Changing stereotypes of the scientist

This study has been built up out of the experiences and opinions of working scientists concerning specialization and change in their careers. No attempt was made to tap their more intimate feelings on this delicate personal subject, nor to place it theoretically in a larger framework of social action. Nevertheless, there emerges from these discussions a general impression of a scientific way of life which is itself changing rapidly in response to wider social and economic changes. This evolutionary process is clearly linked to the transformation of all forms of material human activity − industry, medicine, war, agriculture, etc. − under the direct influence of science itself. It is common to all advanced societies, and transcends national frontiers. Although the individuals who are most affected by this process may not see it as an historical phenomenon in which they are inevitably caught up, they are certainly aware of it through its effects, over the years, on their personal careers.

In retrospect, a career is an accomplished fact, often summed up as stretches of routine life punctuated by episodes of rapid change. Whether these changes led to success or failure, they will seem to have the logical necessity of the situations in which they came about. Because that is what actually happened to *us*, their reality cannot be in doubt.

But in prospect, a career is a mere hypothesis. It is the *project* by which we shape our ends. Because we have not yet lived the experiences that it imagines, we have to create them out of the reported experiences of other people. To some extent we rely upon very specific and localized anecdotes about what has happened to our parents, our relations, our friends, and our colleagues. But because we know that none of these particular people is quite like us in thinking or feeling, we turn to biography and fiction for other models. The attitudes that people adopt towards significant events in their lives are thus strongly influenced by the public stock of *stereotypes* of lives such as their's (§3.3).

The representation of 'change' in the stereotypes of the scientific life is thus a significant influence on the way in which scientists approach and adapt to such change. But it is a hidden influence, implicit in what they say but only occasionally expressly stated. That is, it is primarily a source of some of those indefinable, non-rational motivations (§4.1) which lie too deep in the psyche to be probed by the methods of the present study. Nevertheless, it is instructive to consider some of these stereotypes, and see how far they conform to the realities that we have explored and how they are themselves changing with time.

One of the popular images of the scientist is of a person whose whole life is given over, almost from the beginning, to the pursuit of some particular truth. The archetype is Kepler, puzzling for 25 years over the motion of the planets, or Darwin, spending almost as long building up the evidence for his theory of evolution before publishing *The Origin of Specieses*. One does still come across such cases, as instanced by the following story. It was told by a person whose PhD work had inspired a devotion to a subject where, for socio—technical reasons, it was not possible for them to get a permanent job:

' . . . so really I picked [this establishment] firstly because I had to have some kind of job anyway, secondly because it was [a part of the same R&D organization as the establishment where] I could not go . . . and it might not be too long before [the situation changed] and if I was in [that organization] I might be in a position to move. And thirdly, I suppose, it was a job that I could talk myself into . . . .

'When I arrived here I decided I would do my job here during the day, and then I would do my very best to keep up my research in my own subject during [my spare time]. I worked like hell for about three years, working in the evenings and at the week-ends, spending all my leave [taking part in active research on the subject] . . . . The end seemed to come at a conference some, I suppose, three or four years after I got here, when I could see that I was giving what was a good paper, which was way out in front of the subject, but there was still not going to be any job for me.

'So . . . where was I? I thought that real hard work would somehow get me the job, sooner or later, and that was obviously not going to happen. So then − and I was just hopelessly tired − I decided I would try and develop what I was doing here, with the aim of getting myself into a position in [this establishment] where I could generate my own research, and therefore do something which interested me. Which I then did for the next two or three years, got promoted, and then, as everybody [in this group] is well aware, I started promoting research [in my chosen field] with great enthusiasm. Everybody has been invited to collaborate on projects, left, right and centre, and it's beginning to pay off . . . − the money is beginning to come in for [work related to my subject].

'I've paid the price, in the sense that I haven't obviously been able to be at the top of that field, because I haven't been active with the the people who were working on it, although I kept all my professional connections, and . . . . I am President of [the national learned society in the subject] and all the rest of that kind of thing . . . . If the gamble pays off, I would have had six years of not really working at my best [on the research I am paid to do here] but then will have the rest of my time here working on something that I am passionately interested, and think I am equipped to do, and that would be OK. If the gamble does not come off, of course that would be a pity. But there is an example of total non-versatility, in a sense, and deliberate'.

But this inspirational autobiography, conforming so closely to the traditional stereotype of a scientist as a person dedicated to a transcendental goal, is not at all typical of the research profession in general. When they heard it, the other scientists taking part in the discussion expressed their admiration for this 'superb sort of commitment', and their sympathy and regret for a situation which had produced such an 'extreme case of . . . career mismanagement', but they did not echo it from their own experience. It is exceptional for a professional researcher to develop an affective attachment to a particular specialty at such an early stage in their career that they are quite unwilling to work in any other field.

A more up-to-date and prosaic version of this stereotype is of the scientist as the supreme *expert* on some particular subject. Here, the subject itself is essentially arbitrary. Let it be insect eggs, or helicopter transmissions, or the novels of Jane Austen — it is all the same provided that one gets to know more about it than anybody else in the world. Of course, in time, this relationship of mastery and exclusive possession usually develops into a powerful emotional attachment, but that is not of its essence.

As we have seen, many scientists do follow careers based on this model. Scientific experts are more in demand than ever before. The cultivation of expertise provides a rational motive both for concentration on a subject specialty in the first place (§1.3), and for continued persistence in that specialty (§5.2). Nevertheless, scientists themselves regard this stereotype as slightly opportunistic and worldly — appealing more to the vanity and the pocket than to the heart and the spirit. And acknowledged expertise often turns out to be quicker to achieve, and easier to give up, than is usually realized.

Another variant of the public image is that of the *problem-solver*, the 'honest seeker after truth' (§4.8). The scientist is conceived of as a person of infinite curiosity and ingenuity, whose eye is always being caught by a puzzling circumstance and who cannot rest until it is explained. This is the stereotype associated with the 'academic' mode of research (§1.2), where success in 'the pursuit of knowledge for its own sake' gains the esteem of the scientific community, not to mention other, more material rewards.

This social reinforcement is undoubtedly effective. As somebody said:

'I'm not just being idealistic. I'd rather be known for a measure of success in having done a really neat piece of science, than having had a [professorial] chair, or something like that.'

The reputational careers of many scientists do, indeed, exemplify this stereotype. But notice that it does not in principle demand subject specialization or high technical expertise. It is true that the majority of scientists follow narrowly specialized research trails (§1.4) because they believe that these will lead them to a succession of deep, difficult and interesting puzzles which they will be well prepared to solve (Ziman 1981). But others trust to their native wits to take on any problem that fires their curiosity, over a wide field (§1.5).

The stereotype of the scientist as problem-solver does not, therefore, stand directly in the way of changes in the range of problems to be solved. Nor does it limit the class of problems worthy of attack. As many scientists discover nowadays in the course of their careers (§4.2), solving practical technological problems can be just as satisfying as finding answers to less 'urgent' (§2.3) questions of no immediate relevance. In the day by day work of problem-solving, there usually turns out to be little distinction between 'pure' and 'applied' science. Yet this is a distinction that was much emphasized in the academic tradition. Stress was laid on the freedom of the pure scientist to choose problems for research, by contrast with the applied scientist who was employed to work on prescribed problems. This comes up, for example, in a discussion of the place of commissioned research (§2.4) in research council establishments:

‘ . . . although, in theory, everyone [in this establishmen] . . . does 50per cent applied work, and 50per cent theoretical work in basic research, in fact it doesn't work like that. Some people would do 100per cent basic work, some people would do 100per cent applied. Now, if a wind of change . . . comes about, like the Rothschild Report, or some other outside pressures to link you more closely with the customers, you may have a shorter time scale on the sorts of questions they want to ask. The extent to which you fit in well with that will have to do with how well you fitted in with the old system — and not everybody fits in perfectly. So the pre-conditions for mobility actually exist in terms . . . of what [the establishment] is currently doing . . . . People who will move under these conditions are people who will be forced to do applied work, who either may not be good at it, for a whole variety of reasons. They may not be able to interact with customers and clients very well, or they may not just like that particular type of work. If they are any good, perhaps the question is one of competence. If they are both competent and not very good [*sic*], which is partly why they were so well situated in the first place, they will be the people who will potentially move on.’

But *strategic autonomy* (§3.4) does not seem now to be such an essential feature of this stereotype. Some of the early studies of the management of industrial R&D (Marcson 1960; Orth *et al.* 1965; Pelz & Andrews 1976) indicated that managers needed subtle tactics and considerable time to wean newly recruited PhDs away from their belief that they could only be really happy if they were quite free to work on problems of their own choice. Cotgrove & Box (1970) insisted that the ‘private’ (i.e. non-publishing) scientist working as a professional in the local research context of an industrial firm required just as much autonomy as the dedicated ‘public’ scientist, embracing ‘science’ as a total personal role.

But that was in an era when a person with a PhD could easily get an academic job, and thus had some leverage in demanding terms of employment satisfying this ideal. Nowadays, with the whole research profession under pressure (§3.5), scientists cannot afford to be so fastidious. More recent studies (Bailyn 1981, 1984) suggest that *technical autonomy* alone — freedom to tackle problems in one's own way — is sufficient to ensure personal commitment and self-activation in research (§3.4). The supposed difference between ‘pure’ and ‘applied’ science has turned out to be more of an ideological distinction between competing sub-groups within the R&D community than an essential feature of science as a profession. The replacement of ‘academic freedom’ by ‘technical autonomy’ in this stereotype indicates that scientists are now supposed to be much more biddable and flexible than they used to be in the range of problems they can be expected to solve.

9.4 Role transitions in the career of a QSE

The traditional stereotypes of a career as a scientist (§9.3) had one feature in common: they did not envisage any movement out of research. Scientists who are employed by R&D organizations where these stereotypes are the norm do not find it easy to make such a drastic career change. But these stereotypes are being supplanted by a more general image, which implies much greater personal adaptability. In the public discourse of journalists, politicians, civil servants and businessmen, any *professionally qualified technical worker* is now referred to as a 'scientist', if he or she is in any way involved in R&D − or, even more broadly, in 'RDD&D' − Research, Development, Design and Demonstration. Even in the official occupational statistics, research scientists are seldom differentiated from other 'QSEs' (§3.1).

The careers of many of the people we are concerned with undoubtedly fit into this new stereotype. Quite a number of scientists move from task to task within an extensive domain of activity, often without changing their employers. By the time they reach mid-career, many people who were originally recruited as basic researchers in physics, mathematics, or chemistry (§7.2) have been transformed into technological 'trouble-shooters' or design engineers (§4.3). Others have become involved in the technical administration of their organisation (§6.3), or have taken on its service functions (§6.4). A few have been promoted into managerial positions (§6.2) where they are neither 'experts', nor 'problem solvers', but 'executives', who have to take more general decisions about people and programmes.

The QSE stereotype takes little account of the reputational aspects of a scientific career (§3.3). It is true that scientists in some R&D organizations outside the academic sector are still subject to quasi-academic criteria for promotion (§5.3), but the attainment of high public standing in the scientific community is not the major purpose of their labours, nor the primary goal of their ambitions. In the end, it is their organizational careers that really count (Sofer 1970; Barnes 1971).

Nevertheless, all the professional groups covered by this stereotype do have one characteristic in common: their members usually say that they get immediate satisfaction out of their jobs. Their work means more to them than a means of earning a living, or achieving public success, or providing a service to the community. This emotional attachment to the performance of their craft, this pride in the artistry they must exercise daily (§3.4, §5.5), is continually voiced by the scientists who have contributed to this study. It shows up, for example, in their scorn or sympathy for those who have 'gone stale' — i.e. who can no longer enjoy research as they once did (§5.6).

This enjoyment of the actual technical work of RDD&D goes beyond the vocational commitment (Becker 1960) that is characteristic of most professions. For example, design engineers not only attach a strong symbolic meaning to their occupation, and identify their own careers with its skilful performance (Bailyn & Lynch 1982). They also get direct satisfaction from 'concrete puzzle-solving', suggesting, perhaps, that they share the 'problem-solving' stereotype of the researcher, even though much of their work is of a much more routine nature.

Indeed, as Bailyn (1981) points out, for many QSEs, this can give rise to a 'contradiction . . . between work and career . . . . What is necessary for career advancement is personal exposure in front of management; but effective work requires involvement and responsibility at the working level which characteristically does not include supervisory personnel'. The contradiction here is between the different *roles* each person may be expected to play in the course of his or her career, even within the same organization. The role of the highly specialized researcher or designer is not altogether consistent with the role of the technical administrative assistant, although they are both embedded in the same organizational framework, they are both covered by the stereotyped image of the 'scientist', and they both call for thorough knowledge of a narrow range of subjects.

At any one moment, most of the roles actually being performed by the 'scientists' in an RDD&D organization are highly differentiated and very diverse. But these various roles do not define a corresponding diversity of individual careers. During a working life, a person may move from one such role to another — or even have to carry out several roles simultaneously. This has always been the case with engineers, who start out their careers expecting — even hoping — that their talents and training will carry them from strictly technical roles into managerial positions. In fact, by the age of 50, more than 40per cent of them will have made this transition (Berthoud & Smith 1980). The same applies to most research scientists entering industrial R&D organizations: they soon discover that their careers, also, are likely to lead them away from 'bench' research into other technical or administrative functions. In other words, for the typical QSE in the private sector, changes of *role* are not regarded as *career* changes, but as incidental events along a normal career path.

This is not to say that such careers do not run into the difficulties with which this study is concerned. But as we saw in §7.6, the problems of adaptation to change in mid-career seem to derive less from undue subject specialization than from finding a satisfying and effective personal role within the organization. All organizations employing professional people face this issue, which is much too general to be discussed in detail here. The following remarks are intended simply to indicate what is involved.

One way of comprehending the issue is to see how well individuals accept the different 'value orientations' characteristic of various organizational roles. These orientations can be mapped schematically as a 'field', where four major positions can be defined (Thomason 1970):

'1. *Professional*: Success is measured in terms of contribution to man's knowledge and depends upon assessment by other professionals . . . .

2. *Communal*: Success is measured in terms of achieving a satisfactory extra-work relationship of a relatively localized nature . . . .

3. *Managerial*: Success is measured in terms of achieving high position in a hierarchical organization, but not necessarily the present one . . . .

4. *Organizational*: Success is measured in terms of achievement of high position or reward in the present organization, to which the individual has developed a loyalty . . . .

This typology is obviously very schematic, but it does cover the various non-research functions discussed in Chapter 6. Recent work by Bailyn (1980), directed specifically at the organizational roles of QSEs in mid-career, suggests a more direct categorization, where 'Organizational Evaluation of Effectiveness' also plays a part.

Orientataion at mid-career		Organizational evaluation of effectiveness		
		High (+)		Low (ordinary) (−)
Technical (T)	T +	Independent contributor Policy specialist 'Idea innovator' 'Internal entrepreneur	T −	Technical support Expert on 'formatted' task 'Master'
Human (H)	H +	Top management Sponsor Development as policy 'Successful manager'	H −	Mentor Individual development 'Coach' 'Effective manager'
Non-Work (NW)	NW +	Specialist Internal consultant 'Variance sensor' 'Scanner'	NW −	Routine tasks Reduced time commitment

As Bailyn thus indicates, positive organizational roles can be found for relatively mediocre people, such as those in the 'H-' group, who would otherwise tend to be regarded as 'dead wood', and those in the 'T-' group, who easily become disaffected if they are not given jobs that challenge their technical interests. This table also suggests that it is a mistake to underestimate the organizational contribution of employees with a 'non-work' orientation – i.e. those who are more involved with their families and other personal activities than they are with their work. Although withdrawal into private life is often a symptom of career stress (§5.8), there are still many organizational roles that can be performed perfectly correctly without the high degree of personal commitment required in responsible management or in 'self-winding' research.

In the present study, however, we are primarily concerned with those scientists – perhaps the majority – who do not, in fact, move out of active research in mid-career. Yet even a simple change of research specialty may have some of the features of a more radical role transition. Thus, for example, a scientist may have developed such a close personal identification with a particular subject (§3.4) that he or she may suffer a loss of 'self-concept' comparable to the loss that afflicts many workers when they retire from a regular job (Kosloski *et al.* 1984). For the person concerned, it may count as a major career transition, and thus as one of those life events which are 'for the most part, examples of personal dramas in which esteem valuations are central' (Sarbin 1984). In other words, some of the problems that arise in such a change are worthy of study not only as features of a significant career transition but also from the parallel perspective of the literature on the adaptation of individuals to life events (Sokol & Louis 1984).

In the majority of cases, however, highly charged affective factors do not prevent people from adjusting to change, whether in their working conditions or in their private lives. The real question is whether the *mode* of adjustment is consonant with their longer-term aspirations and responsibilities. But because research is not a routine activity (§3.4), the work role of the scientist requires high levels of novelty within a framework of high personal discretion. As Nicholson (1984) points out in his general theory of work role transitions, these particular requirements can only be met in the mode of 'exploration', where there is considerable development both of the person making the transition and of the role into which he or she moves. He further argues that such a mode of transition is facilitated by prior occupational socialization into roles with similar characteristics, and by a strong desire for personal control and feedback in the new role. In broad terms, these are indeed the conditions under which scientists find that they can adapt successfully to changes of research specialty in mid-career.

Although Nicholson's theory is very schematic, it could obviously be used as a general conceptual framework for the material reported in this book. But personal diversity eventually defies all schematization. The people we are considering are intelligent, socially competent adults, who are alive to their situation in the world and have a well-founded confidence in their own powers. They have spoken up for themselves, and it is well worth listening to what they have to say.

Bibliography

Allen, V. L. & van de Vliert, E. (eds.) 1984 *Role Transitions: Explorations and Explanations* (London: Plenum)

Bailyn, L. 1977 "Involvement and accommodation in technical careers: an inquiry into the relation to work at mid-career" in Van Maanen, J. (ed.) *Organizational Careers: Some New Perspectives* (London: Wiley) . . . §5.8

Bailyn, L. 1980 *Living with Technology* (Cambridge Mass.: MIT Press) . . . §7.1 §9.4

Bailyn, L. 1982 "Resolving contradictions in technical careers" (*Technology Review*, Nov./Dec. 1982, 40−7) . . . §2.5 §3.4 §5.5 §5.6 §7.6 §9.3 §9.4

Bailyn, L. 1985 "Autonomy in the industrial R&D lab" (*Human Resource Management*, **24**, 129−46). . . §3.4 §9.3

Bailyn, L. & Lynch, J. T. 1983 "Engineering as a life-long career: its meaning, its satisfactions, its difficulties" (*J. of Occupational Behaviour* **4**, 263−83) . . . §9.4

Barnes, B. 1971 *Science Studies*, **1**, 157−75 . . . §9.4

Barnes, B. 1985 *About Science* (Oxford: Blackwell) . . . §5.5

Becker, H. S. 1960 *Am. J. Sociol.* **65**, 32−40 . . . §3.4 §5.5 §9.4

Becker, H. S., Geer, B., Hughes, E. C. & Strauss, A. L. 1961 *Boys in White: Student Culture in Medical School* (Chicago: Univ. of Chicago Press) . . . §5.6

Berthoud R. & Smith, D. J. 1980 *The Education, Training and Careers of Professional Engineers* (London: HMSO) . . . §9.4

Bucher, R. & Stelling, G. 1977 *Becoming Professional* (London: Sage) . . . §4.7

Chubin, D. E. 1976 *The Sociological Quarterly*, **17**, 488−96 . . . §1.1 §1.6

Chubin, D. E. & Connolly, T. 1982 "Research trails and science policies: local and extra-local negotiation of scientific work" in Elias, N. Martins, H. & Whiteley, R. (eds.) 1982 *Scientific Establishments and Hierarchies* (Dordrecht: Reidel) 293−311 . . . §1.4 §5.6

CSS 1986 *UK Military R&D* Council for Science and Society (Oxford: Oxford University Press) . . . §2.5

Cotgrove, S. & Box, S. 1970 *Science, Industry & Society* (London: George Allen & Unwin) . . . §9.3

Decker, W. D. & van Atta, C. M. 1973 *Research Management*, **16**, 20−3 . . . §9.1

de Mey, M. 1982 *The Cognitive Paradigm* (Dordrecht: Reidel) . . . §1.1

EASST (European Association for the Study of Science and Technology) 1984 *Newsletter* (Amsterdam: EASST) . . . §2.1

Eiduson, B. T. 1973 "The scientists' personalities" in Eiduson, B. T. & Beckman, L. (eds.) 1973 *Science as a Career Choice* (New York: Russell Sage Foundation) 195−205 . . . §3.4 §4.8

EIRMA (European Industrial Research Management Association) 1970 *The Career of the Research Worker* (Paris: EIRMA) . . . §9.1

Fiorito, J. 1981 *Curriculum Choice and Occupational Attainment in Science and Engineering* (Washington, D.C.: National Science Foundation) . . . §1.3

Fiske, M. D. (chairman) 1979 *The Transition in Physics Doctoral Employment, 1960–1990* (New York: American Physical Society) . . . §7.2

Fox, M. F. 1983 *Soc. Stud. Sci.*, **13**, 283–305 . . . §4.8

Garfield, E. 1979 *Citation Indexing* (New York: Wiley) . . . §1.1

Gieryn, T. F. 1978 *Sociological Inquiry*, **48**, 96–115 . . . §1.1 §1.4 §1.6 §1.7 §5.3

Gieryn, T. F. 1979 ''Patterns in the selection of problems for scientific research: American astronomers 1970–5'' (unpublished PhD dissertation, Columbia University, New York) . . . §1.3 §1.4 §4.8 §5.3 §5.6

Gleave, D. 1985 *The Relationship between Labour Mobility and Technical Change* (London: Technical Change Centre) . . . §5.8

Griffiths, D. 1981 *Personnel Review*, **10**, 4, 14–17 . . . §7.6

Gummet, P. 1980 *Scientists in Whitehall* (Manchester: Manchester University Press) . . . §2.5

Gunz, H. P. & Pearson, A. W. 1977 *R&D Management*, **7**, 173–81 . . . §7.4

Hargens, L. 1975 *Patterns of Scientific Research: A Comparative Analysis of Research in Three Scientific Fields* (Washington D.C.: American Sociological Association) . . . §2.3 §3.4

Hirsh, W. 1981 (private communication) . . . §7.6

Hirsh, W. 1982 ''The postgraduate training of researchers'' in Oldham, G. (ed.) 1982 *The Future of Research* (Guildford: Society for Research in Higher Education) 190–209 . . . §7.2

Holdgate, M. W. (chairman) 1980 *Review of the Scientific Civil Service: Report of a Working Group of the Management Committee for the Science Group (CSD)* (London: HMSO Cmnd 8032) . . . §3.2 §6.3 §7.1 §7.2 §7.6

Hufbauer, K. 1978 ''The role of interspecialty migration in scientific innovation: the case of research on the stellar-energy problem, 1919–39'' (unpublished) . . . §1.6

Hutt, R. 1981 *Policies for Career Change* (Brighton: Institute for Manpower Studies) . . . §3.5 §5.8 §7.1 §7.2 §7.3 §7.4 §9.1

ICSU (International Council of Scientific Unions) 1975 *Physics and Astronomy Classification Scheme* (New York: American Institute of Physics) . . . §1.1

Irvine, J. & Martin, B. R. 1981 *Phys. Tech.*, **12**, 204–12 . . . §4.4 §4.7

Irvine, J. & Martin, B. R. 1982 ''What directions for basic scientific research?'' in Gibbons, M., Gummett, P. & Udgaonkar, B. M. (eds.) 1982 *Science Policy in the 1980s and Beyond* (London: Longman) 67–98 . . . §5.6

Irvine, J. & Martin, B. R. 1984 *Foresight in Science* (London: Francis Pinter) . . . §2.5

IUFRO (International Union of Forestry Research Organizations) 1982 *IUFRO's New Structure* (Vienna: IUFRO) . . . §1.1

Jagtenberg, T. 1983 *The Social Construction of Science* (Dordrecht: Reidel) . . . §2.2 §3.4

Kelly, A. (ed.) 1981 *The Missing Half* (Manchester: Manchester University Press) . . . §3.2

Knight, K. 1976 *J. of Management Studies*, **13**, 111–30 . . . §7.4

Knorr-Cetina, K. D. 1981 *The Manufacture of Knowledge* (Oxford: Pergamon) . . . §1.5 §2.2

Kogan, M. and Henkel, M. 1983 *Government and Research* (London: Heinemann) . . . §2.5

Kosloski, K., Ginsburg, G. & Backman, C. W. 1984 ''Retirement as a process of active role transition'' in Allen & van de Vliert (1984) 331–41 . . . §9.4

Latour, B. & Woolgar, S. 1979 *Laboratory Life: The Social Construction of Scientific Facts* (London: Sage) . . . §2.2 §3.4

Lemaine, G., Lecuyer, B., Gomis, A., & Barthelemy, C. 1972 *Les Voies du Succés* (Paris: Groupe d'Études et de Recherches sur la Science) . . . §2.2 §3.5 §4.2 §5.6 §7.1 §7.2 §7.4 §7.7

MacNabb, G. M. 1981 "A shortage of trained researchers — real or imagined?" (Ottawa: Nat. Sci. & Engin. Res. Council of Canada) . . . §6.5

Marcson, S. L. 1960 *The Scientist in American Industry* (Princeton: Princeton Univ. Press) . . . §9.3

Marshall, W. 1969 *Harwell and Industrial Research* (London: Oxford Univ. Press)

Martin, B. R. & Irvine, J. 1982 "Women in science: the astronomical brain drain" *Women's Studies International Forum*, 5, 41−68 . . . §3.2

Merrison, A. W. (chairman) 1982 *Report of a Joint Working Party on the Support of University Scientific Research* (London: HMSO Cmnd 8567) . . . §3.2

Merton, R. K. 1968 "The Matthew Effect in Science" *Science*, 159, 56−63 . . . §5.6

Morris, J. R. S. (chairman) 1983 *The support given by Research Councils for in-house and university research* (London: Advisory Board of the Research Councils) . . . §5.8

Mulkay, M. 1974 Science Studies, 4, 205−34 . . . §1.6

Mullins, N. C. 1972 *Minerva*, 10, 51−82 . . . §1.4

Nicholson, N. 1984 *Admin. Sci. Quart.*, 29, 172−91 . . . §9.4

Orth, C. D., Bailey, J. C. & Wolek, F. W. 1965 *Administering Research and Development: The Behaviour of Scientists and Engineers in Organizations* (London: Tavistock) . . . §9.3

Pelz, D. C. & Andrews, F. M. 1976 *Scientists in Organizations: Productive Climates for Research and Development* (revised edition) (Ann Arbor: Institute for Social Research, Univ. of Michigan) . . . §7.3 §7.6 §9.1 §9.3

Polanyi, M. 1958 *Personal Knowledge* (London: Routledge & Kegan Paul) . . . §1.1 §4.7

Porter, B. F. 1975 *Nuclear Physics Manpower* (New York: Amer. Instit. of Physics) . . . §1.3

Porter, B. F. 1976 *Optics News*, January, 43−50 . . . §1.3

R&D Review 1984 *Annual Review of Government Funded R&D 1984* (London: HMSO) . . . §3.2

Ravetz, J. R. 1971 *Scientific Knowledge and its Social Problems* (Oxford: Clarendon Press) . . . §3.4

Reif, F. & Strauss, A. 1965 *Social Problems*, 12, 297−311 . . . §5.6 §6.2

Reskin, B. 1977 *Amer. Sociol. Res.*, 42, 491−504 . . . §4.8

Reuter, H., Tripier, P., Aubert, F. & Lahon, D. 1978 *Le Travail de Recherche dans l'Université: Structures et Determinants* (Nanterre: Centre de Documentation et de Recherche en Sciences Sociales) . . . §3.5 §4.7

Sarbin, T. R. 1984 "Role transition as social drama" in Allen & van de Vliert (1984), 21−37 . . . §9.4

Shinn, T. 1980 *Revue française de sociologie*, 21, 3−35 . . . §2.2

Shrum, W. S. 1984 *Soc. Stud. Sci.*, 14, 63−90 . . . §2.1

Sofer, C. 1970 *Men in Mid-career: A Study of British Managers and Technical Specialists* (Cambridge: Cambridge Univ. Press) . . . §3.3 §5.5 §6.2 §9.4

Sokol, M. & Louis, M. R. 1984 "Career transitions and life event adaption: Integrating alternative perspectives on role transitions" in Allen & van de Vliert, (1984), 81−94 . . . §9.4

Thomas, L. E. & Robbins, 1979 *J. of Occupational Psychology*, 52, 177−83 . . . §8.2

Thomason, G. F. 1970 *The Management of Research and Development* (London: Batsford) . . . §4.4 §7.4 §7.6 §9.4

TNO 1980 *The Mobility of Scientific Researchers in The Netherlands* (The Hague: Ministry of Education and Science) . . . §9.1

Weber, M. 1918 "Science as a vocation" in Gerth, H. H. & Mills, C. W. (eds.) 1948 *From Max Weber* (London: Routledge and Kegan Paul) 129–56 . . . §6.2

Whitley, R. 1984 *The Intellectual and Social Organization of the Sciences* (Oxford: Clarendon Press) . . . §3.1 §3.4

Wolff, M. F. 1980 *Research Management* July, 8–10 . . . §6.4 §7.6

Zeldenrust, S. & de Laat, P. B. 1982 (unpublished) . . . §7.4

Ziman, J. M. 1960 "Scientists: Gentlemen or Players?" in Ziman (1981a) 87–92 . . . Introduction

Ziman, J. M. 1974 "Ideas move around inside people" in Ziman (1981a) 259–72 . . . §7.7

Ziman, J. M. 1980 *Teaching and Learning about Science and Society* (Cambridge: Cambridge Univ. Press) . . . §4.3

Ziman, J. M. 1981 "What are the options: social determinants of personal research plans" *Minerva*, **19**, 1–42 . . . §1.4 §1.6 §1.8 §2.1 §2.2 §2.4 §8.2 §9.3

Ziman, J. M. 1981a *Puzzles, Problems and Enigmas* (Cambridge: Cambridge Univ. Press)

Ziman, J. M. 1983 *Proc. Roy. Soc. B*, **219** 1–19 . . . §2.1

Ziman, J. M. 1984 *An Introduction to Science Studies* (Cambridge: Cambridge Univ. Press) . . . §1.1 §1.2 §2.1 §2.2 §3.3 §3.4

Ziman, J. M. 1985 "Pushing back frontiers – or redrawing maps?" in Hägerstrand, T. (ed.) 1985 *The Identification of Progress in Learning* (Cambridge: Cambridge Univ. Press) 1–12 . . . §1.2 §7.1 §7.2 §7.4

Zuckerman, H. 1977 *Scientific Elite: Nobel Laureates in the United States* (New York: The Free Press) . . . §7.3

Index

(prepared by Nicola Kingsley)

organizational career 42–8, 82, 86, 92, 103, 133, 144, 167, 177, 183
overqualified 173

paper, scientific 6, 8, 9, 11, 12, 15, 16, 32, 55, 75, 84, 85, 94, 98, 99, 104, 111, 133, 143, 144, 145, 146, 147, 148, 153, 164, 165, 166, 177, 180
peer review 37, 53, 94, 177
pension rights 101
persistence 10–12, 14, 18, 65, 78, 92–7, 119, 149, 181
  undue 19, 93, 94, 95, 115, 121, 133
personnel 26, 39, 110, 117, 130, 131, 139, 173, 184
PGCE 123
PhD 8, 11, 12, 16, 17, 19, 25, 41, 45, 46, 64, 69, 70, 74, 75, 117, 123, 132, 133, 134, 135, 137, 176, 179, 182
policy, national scientific xvii
political factors 7, 26, 39, 175
polytechnic 30, 35, 42, 45, 61, 70, 118, 122, 170
probation 44, 172
promotion xii, 18, 30, 33, 44, 46, 48, 51, 82, 85, 86, 92, 97, 103, 104, 105, 107, 109, 111, 116, 127, 131, 142, 144–7, 148, 149, 150, 159, 169, 177, 180, 183
PSO xiii, 32, 43, 46, 103, 104, 106, 111, 115, 120, 144, 146, 147, 149, 150, 154
public sector xiii, 32–3, 37, 116
'pure' science: *see* 'basic' science

QSE 41, 42, 55, 127, 183–7

rationality in career decisions 81–2
recognition xi, 18, 45, 50, 85, 86, 93, 110, 133, 143, 144, 150
recruitment 55, 131–6, 139, 161, 169, 172, 176, 182, 183
redeployment 56, 122, 123, 135, 140, 162, 169, 170, 172, 173, 174
redundancy 43, 51, 56, 81, 123, 135, 147, 169, 170, 172
relevance, social 26, 31, 38, 50, 52, 61, 142, 143, 145, 170, 174, 181
reputation 16, 18, 54, 94, 104, 106, 143
reputational career 44, 45, 82, 84–7, 92, 94, 103, 112, 133, 143, 144, 167, 177, 178, 181, 183
research assistant 37, 43, 106, 118, 136
research trail 5, 10, 12, 13, 14, 15, 18, 20, 23, 65, 94, 125, 181
retirement 47, 51, 52, 92, 103, 120, 167, 172, 174
retraining 161–3, 171
Rothschild report 182

sabbatical leave 151, 154
satisfaction 48, 49, 50, 54, 86, 87, 88, 89, 106, 107, 109, 114, 115, 128, 136, 140, 183, 184
scientific assistant 45, 70
scientific officer 32, 45
sector, politico-economic 29–34
security 33, 39, 51, 52, 78, 90, 133, 171, 176
'self-winding' 50, 78, 83, 159, 164, 186
senior scientist 105, 123, 145, 147–51, 173, 178
SERC 31
short-term appointment: *see* temporary appointment
skill 1, 16, 24, 28, 41, 44, 49, 54, 56, 58, 60, 65–76, 78, 90, 92, 93, 107, 108, 109, 113, 114–17, 129, 133, 136, 137, 138, 142, 160, 161, 162, 172, 174, 184
  tacit 1, 28, 73–6
Spain 178
SPATS 111, 112, 113
specialist 4, 28, 72, 99, 113, 115, 136, 150
  life-long 12, 18–21
  scientist as xii, xvii
specialty 1, 70, 94
  boundaries of 4, 5, 7, 16, 18, 142
  change of xvii, 10, 11, 14, 17, 58, 65, 82, 85, 93, 103, 104, 125, 138, 139, 149, 157, 158, 161, 173, 175
  narrow 6, 11, 13, 17, 20, 23, 56, 58, 67, 74, 85, 93, 96, 113, 131, 132, 142, 143, 147, 149, 165, 173, 178
  new 7, 13, 15, 164, 171
  persistence in 11, 12, 121, 181
  personal 7–10, 11
  research xii, xvii, 2, 3, 4, 6, 7–10, 11, 13, 14, 16, 23, 67, 73, 119, 121, 137, 146, 148, 160, 165, 175
  subject xvii, 2, 4, 9, 11, 25, 41, 50, 54, 56, 58, 61, 65, 73, 141, 143, 181
  technical 17, 74, 149, 175
specialization 6, 29, 51, 68, 69, 71, 91, 101, 108, 132, 133, 153, 178
  high degree of xii, 2, 13, 15, 23, 83, 117, 138
  in education 61, 62, 63, 64, 131
  influences on xii
  necessary 1, 2, 56
  of equipment 84
  of skills 24, 92
  rigid xi
SPSO xiii, 43, 46, 85, 86, 106, 148, 149, 150
SRC xi, xii, xv, xvi
stagnation 92, 96, 120, 151, 167, 171, 173
staleness 93, 95, 96, 100, 158, 172, 184